The
INTEGRITY COMPASS

Minneapolis, Minnesota

FIRST EDITION 2026

The Integrity Compass: Leading with Ethics in a Changing World.
Copyright © 2026 by Massoud Amin. All rights reserved.

No part of this publication may be reproduced, stored in a retrieval system, or transmitted, in any form or by any means, electronic, mechanical, photocopying, recording, or otherwise, without the prior written permission of the author.

ISBN: 978-1-962834-69-8
10 9 8 7 6 5 4 3 2 1

Book cover and book design Gary Lindberg

The
INTEGRITY COMPASS

Leading with Ethics in a Changing World

Massoud Amin

Minneapolis, Minnesota

To my daughter, Nahid, who entered
this world and changed mine.

"Wisdom is knowing what to do next; virtue is doing it."

David Starr Jordan

Contents

Preface 1

Introduction 5

Part I: The Terrain 9

Chapter 1: The Ethical Terrain of Technology 11
- *Ethics Under Pressure: An Origin Story* 11
- *From Smart Grids to Ethical Clarity* 12
- *The Changing Terrain* 14
- *From Individual Integrity to System Ethics* 14
- *A Framework for Ethical Decision-Making* 16
- *Learning from Past Tragedies: Cases That Shaped Ethics* 18
- *Legitimacy, Trust, and Infrastructure* 22
- *Responsibility in a Connected World* 25
- *Key Takeaways from Chapter 1* 27

Chapter 2: Emerging Ethical Frontiers –
AI, Autonomy and Complexity 29
- *AI and the Delegation of Decision-Making* 30
- *The Quest for Transparency and Explainability* 34
- *Autonomous Systems and the Ethics of Safety* 37
- *Cascading Risks in Complex Systems* 41
- *Adapting Our Ethical Frameworks* 45
- *Key Takeaways from Chapter 2* 49

Chapter 3: Trust – The Invisible Grid of the New Terrain 51
- *Introduction* 51
- *The Currency of Trust: The Invisible Fuel of Systems* 53
- *The Fragile Thread: Trust and Survival in an Unraveling World.* 57

  The Hidden Grid of the 21st Century 60
  Conclusion: Rebuilding Trust, Rebuilding the Future 63
  Key Takeaways from Chapter 3 66

Part II: The Map 67

Chapter 4: Embracing Failure - The Inventor's Path 69
  A Mirror & a Map: The Power of Stories To Guide Us 69
  The Inventor's Path to Success 71
  Key Takeaways from Chapter 4 79

Chapter 5: Listening to Stakeholders -
A Water Pump Lesson 81
  The Promise of an Innovative Solution 82
  Voices from the Ground 83
  The Fall of the PlayPump Hype 86
  Listening as a Key Leadership Skill 87
  Reflection: Beyond Development 89
  Key Takeaways from Chapter 5 90

Chapter 6: Adapt or Die - Lessons from Kodak vs. Fujifilm 93
  Dominance in the Era of Film 94
  The Digital Disruption 95
  Contrasting Outcomes 99
  What Can We Learn? 101
  Adapting in One's Own Life or Work 103
  Key Takeaways from Chapter 6 104

Chapter 7: Trust in Crisis - The Tylenol Case 107
  The Crisis Unfolds 108
  Guided by a Credo 109
  Rebuilding Trust 111

 Values as a Compass *112*
 Broader Implications of Trust *113*
 Tamper-Proofing and Industry Change *115*
 Key Takeaways from Chapter 7 .... *116*

Chapter 8: Empowerment at Work - The Semco Story **119**
 A Maverick Boss *120*
 The Results: Does It Work? *122*
 Trust and Flexibility *123*
 Wider Impact and Voices of Others *125*
 Lessons in Empowerment and Culture *126*
 Key Takeaways from Chapter 8 .... *128*

Chapter 9: Community Empowerment - Microfinance **131**
 The Grameen Model *133*
 Voices of the Borrowers *134*
 Skeptics and Challenges *135*
 The Ripple Effects *136*
 Adapting and Expanding the Model *138*
 Reflections on Empowerment *139*
 Key Takeaways from Chapter 9 .... *140*

Chapter 10: Use Technology for Good -
The Charity: Water Story **143**
 A New Charity Model *144*
 Impact on the Ground *146*
 Voices of Supporters *147*
 Challenges and Adaptations *149*
 Inspiring a Movement *150*
 Key Takeaways from Chapter 10 .... *151*

Chapter 11: Cross-cultural Collaboration -
The NUMMI Story 153
   Culture Clash 155
   Dramatic Results 158
   Sharing the Lessons 159
   Key Takeaways from Chapter 11 161

Chapter 12: Resilience and Renewal -
Rising Strong from Setbacks 165
   Corporate Comeback: Apple's Near-Death and Revival 166
   Community Resilience: Rebuilding After Disaster in Rwanda 167
   Personal Resilience: Malala's Fight for Education 168
   Synthesis: Building Resilient Organizations and Lives 169
   Key Takeaways from Chapter 12 174

Part III: The Inner Compass 177

Chapter 13: Ethical Practice at Scale 179
   From Principles to Action: The Role of Tools 180
   Decision Trees and Scenario Analysis 182
   Risk Analysis and Proactive Safety Engineering 185
   Stakeholder Mapping and Involvement 188
   Building an Ethical Culture in Organizations 190
   Scaling Ethics Across the Supply Chain and Industry 194
   Making Ethics a Daily Habit 195
   Key Takeaways from Chapter 13 198

Chapter 14: Toward a Culture of Ethical
Systems Leadership 201
   Seeing the Whole System 202
   Moral Clarity and Consistency 203
   Earning and Sustaining Public Trust 205

 Enduring Values in Times of Change 208
 Key Takeaways from Chapter 14 211

Chapter 15 - Letters to Practitioners 213
 Letter to a Young Medical Innovator 213
 Letter to a Young Cybersecurity Professional 218
 Letter to an Engineer Under Pressure 224

Afterword - A Final Call to Action 231

Acknowledgements 235

Author Bio 237

Appendix A - Toolkit 239
 Principles and Heuristics for Ethical Decisions 239
 Questions for Maintaining Integrity 242
 Practice Scenarios and Continuing the Conversation 244

Appendix B - Echoes from Other Minds 249

Appendix C - Decision-Making Checklist 257

Appendix D - Suggested Readings 259

Appendix E - Teaching Guide 261

Appendix F - Discussion Prompts 263

Glossary - The Language of Systems 265

Preface

Why devote an entire book to ethics in engineering, technology, and systems thinking? Because the pace of technological change has never been faster and the consequences of our decisions have never been more far-reaching. Today's engineers and leaders wield tools that can reshape economies, ecosystems and societies in an instant; they design the algorithms that recommend products and parole, the power grids that light our homes, the data platforms that mediate our relationships, and the infrastructures that make our communities resilient or fragile. These responsibilities demand more than technical competence—they demand wisdom, humility and a commitment to justice. Ethics is not a luxury or an afterthought; it is the compass that must guide us through an era of unprecedented complexity and uncertainty.

The material that follows is organized into four main parts, each exploring a different dimension of ethical practice:

- **Part I – The Terrain** traces my personal journey from Iran to the United States and the origins of my own ethical compass. This section explains what the new ethical terrain looks like (with the rise of AI use and the global decline of societal trust) and what's at stake. It explains why now more than ever, an ethical compass is a vital tool.

- **Part II – The Map** examines the professional codes, legal frameworks and research norms that translate values into action. This section explores why trust, resilience, and fairness are central to any modern engineering and technology practice. It includes case studies that examine real-life situations in which individuals, organizations, and leaders made a choice to act ethically or not, to adapt and innovate—or not. Let them serve as guideposts for you on your own journey.

- **Part III – The Inner Compass** provides practical tools and frameworks to help you calibrate your own inner compass. Tools such as risk mapping and scenario analysis are included as well as guidelines on how to create a wider culture of ethics in your organization. Finally, this section presents a series of real letters to students and colleagues confronting dilemmas in medicine, cybersecurity, innovation and leadership. These letters are deeply personal and contextual; they take the reader inside classrooms, boardrooms and emergency rooms to grapple with real-world questions. My hope is that these stories will resonate with readers facing their own crossroads and offer practical strategies for making decisions rooted in conscience and compassion.

- **Appendices** at the end of the book include additional concrete tools: checklists, decision trees, and stakeholder mapping exercises drawn from decades of teaching and consulting. It also includes discussion questions for further reflection and a guide for teachers. The glossary provides a basic introduction to systems thinking.

Throughout this expanded edition, I have tried to preserve the voice and detail of the original book. I have refrained from abridging stories or removing technical context, because nuance matters when grappling with ethical questions. At the same time, I have added this preface to orient readers and to emphasize the themes that run across the chapters: the imperative to speak up, the tension between individual and systemic responsibility, the importance of inclusive innovation, and the power of humility and hope. I believe that a high-quality ethics book is one that balances intellectual rigor with lived experience—one that offers both conceptual tools and human stories. I hope that this expanded edition accomplishes that balance and serves as both a guide and a companion for engineers, technologists, managers, policymakers, students and anyone committed to building a more ethical future.

Thank you for requesting a more comprehensive edition. It is my hope that by integrating the full original text and deepening the discussion, this book will not only inform but also inspire. May it serve as a resource, a mirror and a call to action for everyone who believes that engineering and technological innovation should be guided by empathy, justice and curiosity.

— Massoud Amin

Introduction

In the decades since I began teaching courses on leadership and ethics, the world of technology has exploded in scope and complexity. When I first arrived in the United States as a young engineer, personal computers were novelties, the internet was a nascent research network, and the phrase "cybersecurity" had not yet been coined. Today, when you use your phone, you hold in your hands a device more powerful than the computers that guided Apollo to the moon, and algorithms silently decide what news you read, what job opportunities you see, and even whether you are granted a loan or parole. These advances have brought unprecedented convenience and possibilities—but also new vulnerabilities, inequities, and ethical dilemmas that were unimaginable a generation ago.

My journey began in Iran, before coming to the US in August 1978. Those early experiences taught me that technology and policy are never neutral; they can liberate or oppress, depending on who designs them, who controls them, and who is left out of the conversation. I left my home with a handful of books and a head full of questions, determined to learn how systems, physical and social, really work. Throughout my career, I have been fortunate to lead research on smart grids, to advise governments and utilities on critical infrastructure, and to teach thousands of students around the world. Throughout, my compass has been a simple question: Does this innovation strengthen or weaken the dignity and resilience of those who depend on it?

Much has changed since those early classes. The internet has become the nervous system of commerce and culture. Artificial intelligence systems outperform humans at complex tasks, yet still encode our biases. Climate change strains energy and water systems in ways that demand both engineering ingenuity and humility about our place in nature. Pandemics have reminded us how tightly coupled our fates are. Social movements have exposed deep injustice and called for new models of leadership. At the same time, a new generation of engineers and entrepreneurs brings fresh imagination and moral clarity to these challenges. This generation insists that technology serve the many, not the few; that innovation be inclusive and sustainable; that success be measured not only in profits but in wellbeing and equity.

This primer is a map and a call to action. It distills lessons from my own life as an immigrant, a researcher, a teacher, and a mentor. It weaves together stories from students and colleagues across continents who have confronted ethical crossroads in medicine, cybersecurity, energy, and community building. It introduces concepts from systems science that help us see the hidden feedback and tipping points that shape societies. And it offers practical tools—decision-making frameworks, discussion prompts, and checklists—to help you apply these ideas in your own projects and organizations.

Why ethics now? Because the pace of change has never been

faster, and the stakes have never been higher. New generations are inheriting infrastructure and institutions strained to the breaking point, while also inventing new ones in real time. You yourself might be called upon to make decisions that affect the safety, privacy, and opportunity of millions. Those decisions cannot be outsourced to algorithms or postponed. They require a personal compass calibrated to fairness, transparency, and courage. They need communities of practice that nurture accountability. And they require a willingness to ask hard questions about who benefits, who bears risks, and how we honor those who came before us while building a more just future.

This is not a textbook to memorize; it is a companion for a journey you will travel again and again as the world changes around you. Use it to spark conversations with mentors and peers. Use it to challenge your assumptions and broaden your field of view. Most of all, use it to sharpen that internal compass so that when, not if, you find yourself under pressure, you will know which way is true north.

Part I:
The Terrain

Chapter 1:
The Ethical Terrain of Technology

Ethics Under Pressure: An Origin Story

My own journey began half a world away, rooted in a land of ancient culture and upheaval. I was born in Iran and came of age during turbulent times, including the 1979 revolution that upended my homeland. As a teenager, I left for the United States to pursue my studies, carrying with me the resilience and values of generations. My family's story was one of endurance—ancestors who faced invasion and social upheaval yet "never bowed, only rebuilt." These memories planted a quiet mantra in me: others before me have rebuilt after everything fell apart, so I can too. This legacy of resilience and integrity formed the deep roots of my character, giving me the strength to start anew in a foreign land while holding onto core principles.

One formative experience stands out from my childhood. In the mid-1960s, while visiting a remote village near Tabriz, I encountered a tragic scene. A mother in her early thirties died in childbirth with her baby. There was no clinic, no clean water, no electric power—no chance for them to survive. The dark, isolated village lacked every basic support system that might have saved their lives. Shocked, I asked my parents why such a tragedy had happened. Their answer became seared into my memory: "If they had electricity, everything could have been different. But without electricity, nothing changes. It's as if they live in the dark ages." In that moment, sadness and frustration lit a fire in me. I realized even the most basic human needs remain out of reach without the infrastructure that's often taken for granted. This clarity became my compass: a determination to build systems that protect, connect, and empower people. From that day on, engineering was never just about gadgets or equations tome—it was about justice, about making sure no one else would be left helpless in the dark.

From Smart Grids to Ethical Clarity

Decades later, that early compass guided me through a distinguished career in engineering and technological leadership. I devoted my life to strengthening the security and resilience of critical infrastructures—power grids, water, and communication systems—because I had come to realize they are the backbone of modern civilization. After the September 11, 2001, attacks, I was tasked with directing security research for all North American utilities, pioneering the concept of self-healing "smart" power grids that could prevent blackouts. My visionary work earned me the nickname "the father of the smart grid," highlighting how central my contributions were to modernizing and securing the grid. From designing adaptive defense networks to pushing the frontiers of renewable energy integration, every project was driven by the same mission that took root in that dark village years ago: to use technology to protect and serve people.

Throughout my journey, I have witnessed the double-edged nature of progress. Every new technology brings new vulnerabilities. The massive Northeast Blackout of 2003 and the Colonial Pipeline cyberattack of 2021 underscored how interconnected systems can fail catastrophically and harm millions. These events reinforced my ethical clarity. I saw that resilience isn't about avoiding failure at all costs—it's about adapting, learning, and moving forward even when things fall apart. And I knew that leadership in engineering isn't just about technical expertise—it's about trust, collaboration, and ethics in action. In every war zone I helped stabilize, every boardroom or classroom I led, I found that quiet integrity in daily decisions mattered more than any grand rhetoric.

My personal history as an immigrant-turned-innovator shaped my global perspective over time. I never forgot the villagers living "in the dark ages," or the family mantra of rebuilding with dignity. These gave me an unwavering ethical clarity: technology must remain a servant to humanity's needs, not a master of its fate. No matter how complex the systems I worked on—from smart grids to national security - the principle remained simple. If a solution didn't ultimately better people's lives or if it compromised fundamental values, it wasn't truly progress. This conviction became my North Star. It guided me to speak plainly about potential harm and to insist on transparency and truth even when it was inconvenient. It gave me the courage to pause a project if it threatened safety, and the humility to seek counsel when faced with moral uncertainty. In short, the hardships and lessons of my early life "calibrated" my compass toward integrity, ensuring that as I pushed the frontiers of engineering, I never lost sight of the human beings depending on those systems. This is the story of how a boy who saw a village without power became a leader with a clear ethical vision—a vision I now share with you.

The Changing Terrain

In the twenty-first century, engineering and technological ethics have moved beyond the solitary conscience, making calibrating one's inner compass towards integrity more important than ever. Ethics is no longer solely about one person's integrity in isolation—it now encompasses responsibilities at multiple levels: individual, organizational, and systemic. At the personal level, an engineer's integrity and honesty guide every design decision. At the professional or managerial level, industry codes and standards (like those from civil, mechanical, and electrical engineering societies) commit engineers and companies to hold paramount the safety, health, and welfare of the public. And at the systemic level, we recognize that each innovation plugs into larger networks—power grids, transportation webs, digital platforms—and that our ethical duty extends to the emergent behavior of those networks. In short, the things we design can shape society, so we must consider not just "Is my part done right?" but also "What are the broader ripple effects of this system on people and the planet?"

From Individual Integrity to System Ethics

The concept of engineering and technological ethics has evolved alongside the engineering profession. In the early twentieth century, formal codes of ethics were first established by professional societies in response to tragic failures that shook public faith in engineers. Those early codes focused on personal honesty and competency—for example, avoiding falsification of qualifications or plagiarizing designs. Over time, as engineering projects grew larger and more intertwined with public welfare, ethical codes expanded to emphasize public safety and welfare. By the mid-twentieth century, most engineering codes had adopted the paramountcy of public safety as the first canon, famously stating that "engineers shall hold paramount the safety, health, and welfare of the public." This principle emerged in the wake of disasters like the 1919 Boston Molasses collapse (which killed twenty-one people and prompted licensing laws) and later high-profile cases such as the 1970s Ford Pinto scandal. In the Pinto case, a cost-benefit analysis by Ford valued human life at only $200,000 and concluded paying legal damages for deadly fuel tank fires would be cheaper than a design fix—a calculation that led to hundreds of unnecessary deaths and a public outcry. The ensuing lawsuits (e.g., *Grimshaw v. Ford*) and public anger reinforced that treating safety as optional is ethically unacceptable. It became clear that engineers and companies have a duty not only to obey the law but to prevent harm proactively.

By the late twentieth century, ethical thinking broadened further. Society's expectations grew to include environmental stewardship and social responsibility. Professional codes began to reference sustainability and long-term impacts, not just immediate safety. For instance, today, engineers are encouraged to "strive to comply with the principles of sustainable development" in their work. This shift reflects recognition that engineering decisions can affect future generations and global ecosystems. Now in the twenty-first century, we are entering what might be called the era of "systems ethics." A systems-level ethic means widening our lens from individual components to interconnected consequences. Engineers must ask: How do the policies, incentives, and feedback loops in our technological systems encourage or discourage ethical behavior? What unforeseen harm might cascade from this innovation into society at large? In essence, we still need personal integrity and professional duty. Still, we also need a systemic stewardship mindset—a sense of responsibility for how our engineered systems impact societal resilience, equity, and trust.

A Framework for Ethical Decision-Making

To navigate this broader terrain, it helps to use a structured ethical framework. Throughout this book, we will use a three-level "ethical lens" that aligns with the levels mentioned above. This framework serves as a guiding tool when analyzing any situation:

- **Personal Integrity:** What does my inner moral compass say? This level is about honesty, fairness, and courage in one's own decisions. It asks, "Am I doing what I know to be right, even if it's inconvenient?" It includes virtues like truthfulness in reporting data, refusing to cut corners or falsify results, and being willing to raise concerns. An example at this level would be an engineer refusing to sign off on a safety report that they believe is flawed—or acting from personal conscience.

- **Professional Responsibility:** What are my duties to the public and profession? This encompasses adherence to professional codes of ethics, laws, and standards of care in engineering. It means upholding commitments like protecting public safety, maintaining confidentiality, and continually improving one's competence. It also involves the duty to speak up or blow the whistle when one sees risks being ignored. For instance, the code of American Society of Civil Engineers, like others, urges members to be whistleblowers if necessary to prevent harm.

- **Systemic Stewardship:** This highest level asks engineers to consider complex interactions and downstream effects. It involves questions of sustainability, social justice, and long-term system resilience. An engineer exercising systemic stewardship might ask: If we deploy this new AI-enabled power grid control, what are the potential ripple effects on communities, security, and the environment? Who benefits and who might be left out? How do my choices affect the larger system and society over time? It also means working with other stakeholders—policy makers, communities, other disciplines—to ensure the system as a whole moves in an ethical direction.

This three-level framework (personal, professional, systemic) will recur throughout our discussions. It's a tool to analyze cases: when faced with an ethical dilemma, you can "map" it across these levels. For example, consider a decision about automating a safety-critical process in a chemical plant. Personally, an engineer might feel uneasy (integrity alarm) if the automation could fail unpredictably. Professionally, they have a duty per industry standards to ensure any control system is failsafe and reviewed by independent experts. Systemically, they should think about how this automation interacts with the larger plant and community—could a failure cascade into an environmental disaster? By examining all three levels, we get a more complete ethical picture and can make more responsible choices.

Learning from Past Tragedies: Cases That Shaped Ethics

Ethical principles often emerge from hard lessons. Several landmark engineering failures in history have become case studies that reshaped ethical thinking. We've already noted the Ford Pinto case influencing cost-benefit ethics. Let's look at two other oft-cited cases: the Space Shuttle Challenger disaster and the Tacoma Narrows Bridge collapse. Each illustrates different facets of the ethical terrain.

Case Study: The Challenger Disaster (1986) Speaking Up for Safety

On January 28, 1986, the Space Shuttle *Challenger* broke apart seventy-three seconds after liftoff, killing all seven astronauts on board. The root cause was the failure of an O-ring seal in the solid rocket booster, due to unusually cold launch-day temperatures. What makes Challenger a quintessential ethics case is not merely the technical flaw but the decision-making leading up to the launch. Engineers at NASA and its contractor Morton Thiokol knew the cold could spell catastrophe—in fact, the night before launch, Thiokol engineers warned that low temperatures could trigger a catastrophe due to O-ring vulnerability. One engineer, Roger Boisjoly, pleaded with management to delay the launch, presenting data that the O-rings had failed to seal properly in past cold-weather tests. However, under pressure from NASA management and schedule urgency (the launch had already been delayed), those warnings were overruled. A Morton Thiokol executive infamously told his engineers to "take off your engineering hat and put on your management hat," ultimately recommending launch. The result was the very catastrophe the engineers feared.

The Challenger disaster starkly highlights the ethical duty to speak up in the face of the organizational pressure to stay silent. Boisjoly and colleague Allan McDonald exemplified personal and professional responsibility—they tried to halt a decision they believed was unsafe, even at risk to their careers. Unfortunately, the organizational culture at the time did not empower these voices; in fact, Boisjoly was shunned and suffered professionally after the disaster for his outspokenness. In the aftermath, the Rogers Commission investigation concluded that NASA's decision-making was seriously flawed, citing "go fever," poor communication, and an erosion of engineering safety culture. The tragedy led NASA to implement changes that encourage open dissent and improve risk communication. For the engineering profession at large, Challenger underscored that silence or acquiescence in the face of known risk is an ethical failure. Engineers must feel empowered to hold public safety paramount, even if it means disagreeing with superiors. The case became a teaching story: a reminder that integrity means little unless one dares to act on it, and that organizations must support, not punish, those who raise safety concerns.

Case Study: The Tacoma Narrows Bridge (1940) – Humility and Accountability in Design

On November 7, 1940, the brand-new Tacoma Narrows Bridge in Washington state—then the third-longest suspension bridge in the world—oscillated violently in high winds and eventually collapsed into the water below. Miraculously, no human lives were lost (one dog perished), but the spectacle of the elegant bridge twisting apart (famously captured on film) stunned engineers. The bridge had been designed to be slim and flexible, which made it economical, but wind-induced oscillations were not well understood at the time. Dubbed "Galloping Gertie," the Tacoma Narrows failure became a textbook example of engineering oversight. In ethical terms, it prompted soul-searching about competence and the limits of knowledge. The engineers were competent by the standards of the day, yet nature revealed an unforeseen flaw. This case reinforced an ethical lesson: engineers must acknowledge uncertainty and build in safety margins. It also accelerated research into the aerodynamics of structures and led to more rigorous wind-tunnel testing for bridge designs. While not a case of negligence or malfeasance (unlike Pinto or Challenger), Tacoma Narrows taught humility—an ethical engineer must continuously learn from failures and share those lessons to avoid repeating them. The profession responded by updating design codes for bridges and establishing peer review processes for innovative designs. In the broader sense, the Tacoma Narrows example highlights that ethical responsibility includes accountability for learning from mistakes. When systems fail, the moral duty is not only to fix the immediate problem but to investigate deeply, publicize findings, and improve standards so that future designs are safer. This is how engineering knowledge—and ethics—progresses, often written in the aftermath of failure.

From these cases (and many others like the chemical plant leak in Bhopal 1984, the Chernobyl nuclear accident 1986, or the Deepwater Horizon oil spill 2010), one sees that ethics is not abstract—it lives in concrete consequences. The evolution of ethics has been driven in part by such events. Each disaster expanded our sense of what could go wrong and what should have been done to prevent it. Over time, this has led to stronger safety regulations, better professional guidelines, and a more systemic outlook.

Today, an ethical engineer is expected to be not just a competent designer but also a risk anticipator, a communicator, and sometimes a whistleblower or a public advocate.

Legitimacy, Trust, and Infrastructure

Ethics in engineering is deeply tied to legitimacy and trust. Our infrastructures—whether bridges, power grids, water supply, or digital networks—are the backbone of modern civilization. The public grants engineers and technologists a certain authority to build and maintain these systems. In return, there is an expectation that these systems will be reliable, safe, and operated in the public's best interest. When that expectation is met, it builds legitimacy: people trust the bridge will not fall, the lights will come on at the flip of a switch, and hidden engineering choices will not compromise their data or health. But when engineering failures or unethical decisions undermine a system's reliability or fairness, public trust erodes quickly—and with it, the perceived legitimacy of the system and its stewards.

Consider the 2014–2015 Flint, Michigan water crisis. In an effort to cut costs, officials switched the city's water source and failed to implement proper corrosion controls. This caused lead from old pipes to leach into the water supply, exposing thousands (including children) to lead poisoning. Engineers and authorities initially dismissed or downplayed residents' complaints about discolored, foul water. It was later revealed that some officials falsified reports and ignored evidence of contamination. The result was a complete collapse of trust: by the time the crisis was acknowledged, the public did not trust the city's water engineers or the government at all. As one investigator put it, "It was such a betrayal of public trust that it really shook me to my core."[1] The Flint case illustrates how responsibility and trust are intertwined: the engineers and officials had a basic commitment to ensure safe drinking water—a duty they disastrously failed to fulfill. This failure wasn't just technical; it was a moral failing toward the community. Once trust was broken, restoring it required enormous effort (replacement of infrastructure, health interventions, legal accountability), and even years later, trust in the water supply remains fragile.

Public trust is not only about safety, but also about fairness and transparency. If a transit system or algorithmic platform is perceived as biased or serving only a few at the expense of many, people lose faith in it. Legitimacy means the public believes a system or technology has the right to operate in society—and that legitimacy is earned by acting in transparent, accountable ways. For instance, the legitimacy of an electric power grid depends on more than delivering electricity cheaply. It also depends on the public's belief that the grid is secure (not prone to massive blackouts), that it's maintained ethically (no negligence leading to disasters), and that it's expanded or regulated with the public good in mind (e.g., not polluting neighborhoods without recourse). When legitimacy fails, collapse follows, as one expert succinctly noted. We can draw an analogy: trust and legitimacy are to systems what structural integrity is to bridges. A bridge might stand for years, but if hidden corrosion is ignored, it may collapse one day. Similarly, a system can function, but if unethical practices corrode trust, the public

---

1 "Flint water crisis: For young engineers, a lesson in the importance of listening," U.S. National Science Foundation, March 23, 2016, https://www.nsf.gov/news/flint-water-crisis-young-engineers-lesson.

support for that system might collapse suddenly—through protest, non-use, or demands for shutdown.

Modern infrastructure crises underscore this. The 2003 Northeast Blackout started from a single utility's equipment problem and a failed alarm system, but cascaded into a multi-state power outage affecting fifty million people. Beyond the immediate economic losses, it shook people's confidence: How could the mighty North American grid go dark in minutes? Studies after the blackout showed that the grid operators had ignored alarm system errors and failed to coordinate with each other, as well as human and organizational factors. In turn, regulators and the public pressed for reforms in grid reliability standards. The lesson resonates: a cascading technical failure can trigger a cascading loss of public trust. As I myself have observed from decades of infrastructure work, a blackout spreads in minutes; restoring confidence in the grid takes years. Trust is far easier to break than to rebuild.

Another example can be drawn from the COVID-19 pandemic and its impact on public health infrastructure. The rapid development of vaccines was an engineering and scientific triumph. Yet uneven communication and historical injustices led some communities to distrust the vaccine rollout. In some cases, technically sound systems faltered because they hadn't earned legitimacy in the eyes of those they meant to serve. Engineers working on medical devices or health algorithms have learned that technical excellence alone doesn't automatically earn public trust; one must engage with communities, ensure equity, and operate with transparency to achieve acceptance.

In essence, infrastructure is not just steel and concrete or code—it's a social contract. The legitimacy of that contract rests on ethical stewardship. Engineers and leaders are custodians of public trust: every time a decision is made to approve a marginal design, delay maintenance, obscure a flaw, or prioritize profit over safety, that trust is either upheld or undermined. Conversely, when engineers take responsible actions—for example, a company issuing a prompt recall for a defective product before any tragedy occurs—they strengthen the public trust that is the foundation of their license to operate.

Finally, it is important to note that legitimacy and trust apply internally as well. Within organizations, engineers must trust that if they raise a concern, management will listen (and not retaliate). That internal trust—a culture of ethics—is what enables companies to catch problems early and avoid betraying the external public trust. If internal trust breaks (say, engineers feel management only cares about profit, or worse, instructs them to hide issues), then whistleblowing or leaks might become the last resort, airing issues in the court of public opinion. In either case, the best outcome is to prevent problems, not offer post hoc apologies. Thus, cultivating trustworthiness inside and out is part of the ethical terrain.

Responsibility in a Connected World

The terrain of modern engineering ethics is expansive and often challenging to navigate. Engineers are no longer just problem solvers; we are also societal stakeholders entrusted with the very frameworks of daily life. Electricity, clean water, safe buildings, efficient transportation, digital communication—these all require diligent oversight grounded in ethical responsibility. With systems growing more complex and global, the potential impact of ethical lapses grows too. A mistake or shortcut in a single component (a line of code, a bolt, a sensor calibration) can ripple out through networks to affect millions. On the flip side, an ethical choice—such as designing a system with an extra fail-safe or taking the time to consult community input—can avert disasters and yield widespread benefits.

As you progress through this book and your career, remember that ethics operates on multiple levels simultaneously. In any given project, ask yourself: What is my personal stance and intuition on this? What do my professional standards say? And how does this fit into the bigger human and environmental picture? These questions map to the personal, experienced, and systemic dimensions we discussed. By keeping all three in view, you greatly increase the likelihood of making choices that stand the test of time.

Engineering is often about trade-offs—cost vs. performance, speed vs. safety, convenience vs. privacy. Ethical thinking doesn't remove the need to make hard trade-offs, but it ensures that values like safety, honesty, fairness, and sustainability are given their due weight in those decisions. It widens the definition of success beyond just technical function or profit to include the human consequences. In doing so, it guides us to designs and decisions that earn trust and deliver lasting value.

In summary, the ethical terrain of modern engineering is defined by broad horizons and interlinked paths: from the inner compass of individual integrity and resilience, along the well-marked trails of professional duty, out to the open vistas of systemic impact. It is a terrain marked by historical landmarks—the cases and codes that came before—and new frontiers that continue to emerge. Navigating it requires both a map (frameworks and lessons) and a compass (one's values). In the following chapters, we turn to some of those new frontiers reshaping the landscape, such as artificial intelligence and autonomous systems, which are testing our ethical maps and compasses in unprecedented ways. But as we do so, keep in mind the core insight of this chapter: engineering ethics is ultimately about maintaining the trust and well-being of society. Legitimacy, once lost, is hard to regain; responsibility shirked, once harm occurs, is hard to forgive. Thus, we carry forward the charge that every generation of engineers has had—to be worthy of the public's trust—but we do so in a context that is broader and more complex than ever before. The terrain is challenging, but with careful study and sincere intent, we can travel it honorably and leave behind a firmer road for those who follow.

Key Takeaways from Chapter 1

- **Ethics on Three Levels:** Modern engineers must think ethically as individuals (personal integrity), as professionals (adhering to codes and duty to the public), and as system stewards (considering broader societal and environmental impacts). Using this three-level framework ensures a comprehensive view of our responsibilities.

- **Evolution of Ethical Standards:** Engineering ethics has evolved from simple rules of conduct to proactive duties. Historical tragedies taught hard lessons that transformed industry standards—for example, prioritizing safety above all and empowering engineers to speak up when risks loom.

- **Learning from Failures:** Ethical lapses—whether ignoring an engineer's warning or compromising on public health—can have dire consequences. Each failure underscores specific ethical principles: the duty to speak out, the necessity of competence and humility, and the obligation to protect the vulnerable. Wise engineers treat past failures as lessons and strive not to repeat them.

- **Trust and Legitimacy:** Public trust is the invisible foundation of all infrastructure. When engineers do the right thing, they strengthen the legitimacy of technology in society. When they act unethically or neglect safety, they betray public trust, often sparking outrage and loss of confidence. Ethical engineering is essential to maintaining the social license to operate.

- **The Ethical Compass:** Ultimately, navigating the ethical terrain requires both frameworks and moral conviction. Codes, laws, and analysis tools provide guidance (the map), but an engineer's personal values and courage (the compass) determine how well they follow that map, especially under pressure. Keeping the welfare of people and the planet as the true north of that compass will help ensure engineering work remains a noble and trusted endeavor.

Chapter 2:
Emerging Ethical Frontiers – AI, Autonomy and Complexity

Why does considering ethics matter? Because rapid technological advances are reshaping the ethical landscape. In particular, artificial intelligence (AI), autonomous systems, and ever-more complex cyber-physical infrastructures are presenting novel dilemmas that our traditional ethical frameworks are struggling to address. We now face questions that sound like science fiction but are all too real: Can an algorithm be prejudiced? Who is responsible if a self-driving car causes harm? How do we maintain human oversight when decisions are made in microseconds by machine networks? This chapter explores these emerging ethical frontiers. We will examine the challenges posed by delegating decision-making to machines, the push for transparency and fairness in algorithms, and the systemic risks that arise when technologies interconnect on a vast scale. Through real-world case studies, we will see how these issues have played out in recent years—sometimes with troubling consequences—and discuss how engineers and society are grappling with the trade-offs.

AI and the Delegation of Decision-Making

Artificial Intelligence systems have rapidly permeated many domains of life, from the mundane (recommending movies or filtering email spam) to the momentous (screening job applicants, diagnosing diseases, controlling power grids). Delegating decision-making to AI offers immense benefits in efficiency and consistency, but it also raises a fundamental ethical concern: *we are ceding a degree of agency to machines.* When an algorithm decides who gets a loan, or which neighborhood is targeted on police patrols, or how a vehicle reacts to an impending collision, those are decisions with ethical weight. Yet an algorithm has no moral sense—it optimizes whatever it was programmed (or trained) to optimize. Thus, the onus falls on the human designers and deployers of AI to ensure those decisions align with our values and do not cause unjust harm.

One major challenge in this delegation is algorithmic bias and fairness. AI systems, especially those based on machine learning, learn from historical data. If that data reflects societal biases or inequalities, the AI can inadvertently amplify those biases. This has been well documented: for example, facial recognition algorithms have shown significantly higher error rates for people of color and women than for white males. A landmark 2018 study by researchers Joy Buolamwini and Timnit Gebru found that some commercial facial analysis algorithms had error rates of 0.8% for light-skinned men but 34% for dark-skinned women.[2] Such disparities led to multiple incidents where innocent Black men were wrongfully ar-

---

[2] Rachel Fergus, "Biased Technology: The Automated Discrimination of Facial Recognition," ACLU Minnesota, February 29, 2024, https://www.aclu-mn.org/news/biased-technology-automated-discrimination-facial-recognition/.

rested due to false face recognition matches—essentially accused by a biased machine. In another case, a major technology company had to scrap its AI recruiting tool when it was discovered that the system was downgrading résumés with female indicators because it had been trained on data from a male-dominated tech industry. The tool learned from past biased hiring decisions and, as a result, reproduced the bias by favoring male candidates. These examples show how AI can bake in and even magnify human prejudices under a veneer of objectivity.

The ethical issues here are multi-fold. First, there is the issue of justice: individuals can be denied opportunities or freedoms (jobs, loans, parole, etc.) not for any legitimate reason, but due to a skewed algorithm. This undermines the principle of fairness and can reinforce historical discrimination. Second, there is a concern about transparency and accountability: AI decisions can be opaque, especially with complex models like deep neural networks. If you are denied a loan by a human loan officer, you might at least get an explanation ("your income is below the threshold"); if an AI denies you, often neither you nor the company's own staff fully know why the model made that prediction. This "black box" nature makes it hard to challenge or appeal decisions, raising issues of due process and autonomy for those affected.

Society is grappling with how to rein in these problems. There is now a field of algorithmic fairness and AI ethics devoted to developing techniques for bias detection and mitigation in AI. For instance, tools can test an AI model with diverse inputs to see if it yields disparate outcomes for different demographic groups. Regulatory efforts are also underway: the European Union's GDPR includes a right to explanation for automated decisions, and some jurisdictions have started requiring algorithmic impact assessments for high-stakes uses of AI (similar to environmental impact assessments for big projects). Companies like IBM, Microsoft, and Google have announced principles for ethical AI, emphasizing values like fairness, transparency, and inclusivity. Yet, implementing these values is challenging. One practically effective approach has been to keep humans "in the loop" for critical decisions—meaning AI provides a recommendation but a human decision-maker must review and approve it, especially if the decision negatively affects someone's rights. This is used in some settings like medicine (AI may flag abnormal scans, but a doctor diagnoses) or criminal justice (risk assessment algorithms might inform a judge, but not replace judgment). However, even human oversight can fail if the human is too deferential to the algorithm's seeming objectivity—a phenomenon known as "automation bias."

Case Study: Algorithmic Bias in Policing

A telling real-world case is the use of predictive policing algorithms by law enforcement. These tools analyze crime data to predict where crimes are likely to occur or who might reoffend, to allocate police resources more effectively. In theory, this sounds objective – "data-driven policing." In practice, it has raised serious ethical red flags. Investigations found that systems like PredPol (now called Geolitica) would often send police disproportionately to neighborhoods that had historically been over-policed (usually poor, minority communities), not necessarily because those areas had more actual crime, but because they had more recorded crime data—a classic feedback loop. One analysis noted that predictive policing can create a self-perpetuating cycle of prejudice: because minority neighborhoods have more police and more arrests historically (often for minor infractions that go overlooked elsewhere), the algorithm predicts more crime there, leading to heavier policing, leading to even more recorded incidents, and so on. Meanwhile, crimes in affluent areas (drug use behind closed doors, for example) go under-reported, and thus the algorithm deems those areas "low risk," further diverting patrols away.

In effect, the technology ends up reinforcing existing biases under the guise of neutrality. The ethical implications are profound—such tools, if unchecked, can unfairly burden certain communities and undermine trust in law enforcement. Indeed, when it came to light, public outcry and civil rights scrutiny led some cities to suspend predictive policing programs, and some U.S. Senators called for halting their use due to evidence that "they worsen the unequal treatment of Americans of color by law enforcement." This case reinforces that delegating decisions to AI does not absolve us of moral responsibility. Engineers and officials must critically examine the data and assumptions behind AI systems and be willing to pull the plug if they clash with societal values of equality and justice.

The Quest for Transparency and Explainability

A key ethical frontier with AI is decision transparency—making machine decisions understandable to humans. Opaque AI systems not only frustrate those affected, but they also pose a risk in safety-critical or rights-critical applications. Imagine an autonomous medical diagnosis system that recommends a particular treatment; a doctor would rightly want to know why that recommendation was made, especially if it contradicts the doctor's own judgment. Or consider an AI that flags certain financial transactions as fraudulent; banks and customers want to ensure it isn't arbitrarily denying service or inadvertently redlining certain groups.

The field of AI explainability has emerged to tackle this. Researchers develop techniques like surrogate models (simpler approximations of complex models), attention visualization (showing which inputs influenced the AI's decision), or rule extraction (deriving human-readable rules from the machine's behavior). These can help provide at least partial explanations. For instance, an explainable AI system in healthcare might highlight that "the algorithm flagged this patient as high-risk because of factors X, Y, Z in their lab results and medical history," allowing a physician to review those factors.

From an ethics perspective, explainability is tied to accountability and informed consent. If people cannot understand an AI's workings at all, it's difficult to hold anyone accountable for bad outcomes—because responsibility can be deflected ("the computer said so.") We have already seen legal debates over who is liable when AI causes harm (e.g., if a self-driving car makes a mistake, is it the manufacturer's fault? The software developer's? The passenger's for trusting it?). Clear explanations can assign responsibility more fairly. Moreover, for individuals whose lives are affected by AI decisions, some level of explanation is important to preserve dignity and autonomy. Being subjected to a mysterious algorithmic judgment with no explanation can make people feel powerless. Conversely, providing reasons enables understanding and leaves room for potential contestation—aligning with principles of due process.

Case Study: The "Black Box" of Credit Algorithms

A real-world example: In 2019, some customers of a new credit card (the Apple Card, managed by Goldman Sachs) noticed a troubling pattern—women were being given significantly lower credit limits than their male spouses, even when the women had higher incomes or better credit scores. This came to light when a tech entrepreneur tweeted that despite his wife having a better credit score, her Apple Card limit was one-tenth of his. Similar reports followed, including one from Apple's own co-founder, Steve Wozniak, whose wife also got a far lower limit. Naturally, people asked the bank for explanations. The bank insisted the algorithm used no gender information and blamed it on a "black box." The lack of transparency made it impossible to know if there was algorithmic bias at work (indirectly through variables correlated with gender) or some other factor. The incident spurred an investigation by regulators in New York. Goldman Sachs eventually apologized and adjusted some limits but notably stated they could not share the exact algorithmic reasoning due to its complex and proprietary nature. This case underscored the frustration and suspicion that opacity breeds. Even if there was no intentional discrimination, the inability to explain the logic behind the decision damaged trust in the product. Many saw it as a cautionary tale that if you can't explain your algorithm's outcomes, you shouldn't be using it for decisions that have serious personal impacts.

Another area where there's a call for transparency is AI in criminal justice. For example, risk assessment tools that predict a defendant's likelihood of reoffending are sometimes used in sentencing or bail decisions. One such tool, COMPAS, was challenged when an investigative report found it often gave higher risk scores to Black defendants than white ones for equivalent profiles. The company that developed COMPAS kept its algorithm secret (proprietary), which meant defendants couldn't challenge how their score was determined—a due process concern. Critics argued this was akin to "algorithmic secrecy" undermining justice, and some courts have since ruled that if such tools are used, the defense has a right to examine their workings. The broader principle emerging here is that algorithmic decisions with significant consequences should be subject to explanation and, where appropriate, independent audit. Black boxes may be tolerable for, say, recommending songs, but not for decisions affecting liberty, livelihood, or life.

Autonomous Systems and the Ethics of Safety

We are also witnessing the rise of autonomous machines—self-driving cars, drones, autonomous robots in warehouses and hospitals, even autonomous weapons. These systems move and act in the physical world, often using AI for perception and control. They carry the promise of great benefits: autonomous vehicles (AVs) could reduce accidents (since human error causes most crashes), drones can aid in disaster relief, and robots can perform dull or dangerous tasks. Yet they also introduce new ethical dilemmas and safety questions.

One immediate ethical concern is safety and liability: what happens when an autonomous system fails? This is not hypothetical—there have been multiple accidents involving self-driving cars in testing. In 2018, a self-driving Uber test vehicle tragically struck and killed a pedestrian, Elaine Herzberg, in Tempe, Arizona. Investigations found that the car's sensors did detect the woman pushing a bicycle across the road in the dark, about five to six seconds before impact. However, the system's AI struggled to classify what it was "seeing" —initially tagging her as an unknown object, then as a bicycle, then something else—and it never correctly predicted her path. The emergency braking system was turned off (to avoid false alarms), relying on the human safety driver to intervene. Unfortunately, the safety driver was momentarily distracted (reportedly streaming a video on her phone). The vehicle did not issue an alert in time, and the driver looked up a split second too late to brake. This case starkly showed the ambiguity in responsibility with autonomous systems. The company (Uber) had decided on certain system settings and testing protocols, the software had decision flaws, and the human overseer failed to monitor diligently. In the legal aftermath, the safety driver was charged with negligent homicide (she later pleaded guilty to endangerment, receiving probation), whereas Uber as a company faced no criminal charges. Ethically, many asked: was it fair to pin blame mostly on the operator, given the technology and corporate decisions set her up to fail? This problem of handing off responsibility – when humans are supervising automation – is known to be tricky, because highly autonomous systems can lull humans into inattention, yet still expect them to snap to attention in an instant when an emergency arises. It can be argued that Uber's testing program did not adequately ensure safety (e.g., the decision to turn off automatic braking and rely on human intervention was questionable). More broadly, the incident highlighted how accountability can fall into grey zones with autonomy. If a human driver causes a fatal crash, we usually blame the driver. If a fully autonomous system causes one, do we blame the manufacturer? The programmer? The owner? Society is still establishing the rules here.

Another ethical facet of autonomous vehicles is the classic "trolley problem" scenario: if an Autonomous Vehicle must choose between two bad outcomes (say, hitting one pedestrian vs. swerving and risking its passenger's life), how should it be programmed to decide? Human drivers make such split-second judgments intuitively (and legally, any outcome is typically just "accidental"), but an AV might literally be following a preset decision algorithm. Should it minimize total harm (even if that sacrifices its owner)? Should it prioritize the occupants who trusted it? Different people and cultures have different intuitions, as shown by surveys. Manufacturers have mostly avoided explicitly programming trolley-problem logic, instead focusing on doing everything to avoid collision. But implicitly, design choices might still favor one outcome over another (e.g., an AV might be programmed to prioritize staying in-lane no matter what, which could mean hitting whatever is ahead rather than swerving). Ethicists argue that these choices should be transparent and perhaps even regulated, rather than left to secret corporate code.

Case Study: Boeing 737 MAX – Automation and Opaque Design

While not usually thought of as "autonomous" (airplanes still have pilots), the Boeing 737 MAX crisis in 2018-2019 serves as a cautionary tale about automated systems and transparency. Boeing had introduced a new automated anti-stall feature (MCAS) that could push the plane's nose down if it sensed a high angle of attack. This system was not well communicated to airlines or pilots—it was omitted from manuals to minimize training requirements. In two tragic crashes (Lion Air 610 and Ethiopian Airlines 302), a faulty sensor fed erroneous data to MCAS, causing it to push the nose down repeatedly; the pilots, unaware of MCAS's existence or behavior, fought against the plane's automation until they lost control, and both flights crashed, killing 346 people. The ethical issues were manifold: lack of transparency (pilots weren't informed of a critical system), corner-cutting in design (MCAS relied on a single sensor and had authority to override pilot input repeatedly), and corporate pressure (internal documents later revealed Boeing was keen to avoid costly pilot retraining, influencing their decisions).

This case underscores that incremental automation, if not done with great care, can be as dangerous as full autonomy. It also shows the ethical imperative of honesty and completeness when introducing automated features—users (in this case, pilots) must know what the system can do, or else they cannot respond appropriately. Boeing's failure to do so was a breach of trust with those who operate its systems and with the flying public. The aftermath led to the grounding of all 737 MAX planes, reputational damage to Boeing, and an industry reckoning about how such technology should be vetted and disclosed. From an ethical systems perspective, the Boeing case illustrates how organizational decisions (not just technical errors) around autonomy can lead to catastrophe—reinforcing that safety culture and transparency are as important as the code and hardware.

Cascading Risks in Complex Systems

Our modern infrastructures are deeply interconnected. Power grids communicate with internet networks; supply chains span continents on digital platforms; and cities depend on electricity, water, and communications systems that all depend on each other. This complexity brings efficiency, but it also means small failures can quickly cascade into large-scale crises—a phenomenon sometimes called "cascading failure" or "systemic risk." These cascading failures are an ethical concern because they challenge our traditional methods of managing safety (which often assume isolated failures) and because they can affect huge numbers of people who had no part in the precipitating cause. An engineer working on one piece of a system might inadvertently set the stage for a domino effect.

One vivid example was the 2010 Flash Crash in financial markets. On May 6, 2010, the U.S. stock market plunged about 9% within minutes—wiping out nearly $1 trillion in value—then mostly recovered within half an hour. This chaos was driven by high-speed trading algorithms that, in essence, started rapidly selling in response to each other's actions, amplifying a minor market blip into a major crash. It was an unprecedented event showing how the interactions of algorithms (each following its programmed logic) could create a feedback loop beyond any human's anticipation. While markets have circuit breakers and eventually correct, the incident raised alarms about the fragility of automated, interconnected systems. Ethically, it signaled that designers of financial algorithms have a responsibility not just to their firm but to the stability of the whole market. This duty wasn't clearly codified at the time. Regulators have since tweaked rules to prevent a repeat (like trading pauses on excessive volatility), but the risk remains that tightly coupled algorithms could outstrip our control.

Another example is the electric power grid. The aforementioned Northeast Blackout of 2003 started with a tree branch in Ohio taking down a power line. That incident alone should have been manageable, but a combination of a software bug in the grid control room's alarm system (which caused operators to miss early signs of trouble) and the physics of power load flows led to a domino effect: as one line went down, its load shifted to others which overloaded and tripped, causing generators and lines across several states to cascade off for self-protection. Within about nine seconds, the cascade swept from Ohio across Michigan, Ontario, New York, and beyond. Fifty-five million people lost power for hours or days, in some cases with significant economic and public health impacts (e.g., water systems without backup power, traffic chaos, etc.). This event showed how complex system failures can propagate incredibly fast, faster than any human operator can react. It became clear that more automated protection and better real-time monitoring were needed. Still, coordination and shared responsibility across organizations were crucial—no single utility or engineer could see the whole picture that day. In the ethical sense, it highlighted a systemic responsibility: utilities and grid operators must work together and share information for the good of the larger public, rather than think in silos. It also spurred investment in what I championed as "self-healing grids," which use smarter networks and AI to isolate disturbances and reroute power autonomously to prevent wide blackouts. However, as we connect grids with telecom and automation, new risks appear: what if the very systems meant to stop a cascade (like AI controllers) malfunction or are tampered with? This complexity gives rise to cybersecurity ethics—protecting infrastructure from malicious hacking is now part of an engineer's mandate, since a hack can be as destructive as a hurricane. Indeed, the 2021 Colonial Pipeline cyberattack—where hackers shut down a major fuel pipeline via ransomware—caused fuel shortages and panic buying in parts of the Southeast U.S.. A relatively small technical vulnerability in an IT system cascaded into a real-world crisis of gas lines and price spikes. This underscores that systemic risk management is an ethical imperative: engineers must anticipate not just direct failures but also how failures might cascade and how bad actors might exploit the interconnections.

Complexity also raises the issue of unintended consequences. When we introduce a new technology into a complex social-tech system, it may interact with elements in ways we didn't predict. Social media, for example, started as a way to connect friends, but at scale it became a vector for disinformation and societal polarization, affecting democracies. No single engineer at Facebook or Twitter intended that outcome. Still, the emergent behavior of these platforms created ethical quandaries about content moderation, algorithmic amplification of extreme content, and mental health impacts. The cascading effects weren't purely technical; they were socio-technical. Engineers working on such platforms now have to consider issues like: "If we optimize for engagement, do we inadvertently encourage outrage and falsehoods because they grab attention more?" That is a systems ethics question.

How can we manage such complexity ethically? One approach is to adopt a principle of precaution and monitoring. Engineers should be humble about the limits of prediction in complex systems and thus build in monitoring, fail-safes, and iterative checkpoints. For example, when deploying an AI in a critical infrastructure, instead of a full hand-off, there could be a phased introduction with continuous evaluation, allowing for a pullback if anomalies appear. The ethic is ongoing vigilance: ethical responsibility doesn't end at launch, but instead requires tracking how the system behaves in the wild and being willing to intervene if necessary. Another approach is scenario planning—considering ahead of time: What if our trading algorithm interacts with another rogue algorithm? What's the worst that could happen? This is analogous to stress-testing banks or running emergency drills. Engineers have a duty to use their imagination for worst-case scenarios, not because they are likely, but because the harm could be huge if they do occur.

Case Study: COVID-19 Contact Tracing Apps – Balancing Privacy and Efficacy

A contemporary example of a complex socio-technical system was the attempt to use smartphone apps for COVID-19 contact tracing. Technologically, it seemed promising: smartphones could log proximity events via Bluetooth and notify people if they'd been near someone who later tested positive, potentially automating and speeding up contact tracing to contain outbreaks. Several countries and U.S. states launched apps using frameworks by Apple and Google that emphasized privacy (using anonymized Bluetooth tokens, not GPS). The ethical balancing act here was complex: public health benefit vs. individual privacy, and voluntary adoption vs. effectiveness. To respect privacy, these apps were designed to store minimal data and rely on user consent for sharing a positive status—but that very design also limited their effectiveness (few people downloaded them, and those who did were sometimes reluctant to report illness).

In places like Singapore, a similar app was launched, but it later emerged that police accessed the data for unrelated investigations, causing public outcry due to a sense of betrayed trust. In the West, privacy protections were strong, but adoption was low—partly due to a trust issue, ironically. This case shows how even a well-intentioned system can struggle if it doesn't align with social values and trust. The complexity here was not technical (the apps worked as intended technically), but systemic: the success required public buy-in, integration with health systems, and clear communication. The lesson is that introducing new tech in a complex human system requires multidisciplinary ethics: understanding user concerns, legal frameworks, and cultural factors is as important as the code itself.

Adapting Our Ethical Frameworks

The emerging frontiers of AI, autonomy, and complexity do not render traditional ethics obsolete, but they demand adaptation and expansion of our ethical toolkit. For instance, the classical engineering ethic of "ensure safety, test thoroughly" remains crucial, but how do we apply "test thoroughly" to an AI that learns and changes over time or an algorithm that's interacting with millions of unpredictable users? We have to develop new methodologies—like continuous monitoring, bias audits, and maybe even ethics "bug bounties" where outsiders are invited to find ethical flaws in a system (much as security researchers do for cyber vulnerabilities).

Interdisciplinary collaboration is increasingly recognized as vital. Engineers alone can't solve AI ethics or systemic risks; we need to work with data scientists, ethicists, sociologists, lawyers, and the affected communities themselves. For example, to make an algorithm fair, you might need sociologists to point out which metrics reflect social biases, or to craft a definition of fairness that aligns with legal and moral norms. In autonomous vehicle ethics, philosophers and the public might need to weigh in on the value trade-offs (e.g., how cautious vs. aggressive should an AV be, knowing caution might slow traffic but aggression might risk more accidents).

There are also calls for regulatory adaptation. Existing laws on product liability, for example, didn't anticipate AI. We may see frameworks where certain AI systems require explainability by law, or certain high-risk AI (like in medicine or driving) go through certification or inspection, much like engines and bridges do. The European Union's draft AI Act is one early example, proposing to classify AI systems by risk and impose requirements accordingly (e.g., high-risk systems must have documentation, human oversight, etc.). This is analogous to how we regulate medical devices or aircraft—not every gadget needs FAA approval, but a jetliner's flight control software certainly does. Ethical practice will likely involve not just voluntary adherence to principles but also meeting external standards and participating in the conversations that shape them.

Throughout this adaptation, one constant ethical touchstone is the need for humility and caution. The more powerful and inscrutable our creations, the humbler we should be about their potential downsides. This doesn't mean stifling innovation; it means proceeding with eyes open and safeguards in place. In complex systems, we must respect that we might miss something. So an ethical engineer will advocate for "graceful failure" mechanisms—designs that err on the side of safety when they err. For a self-driving car, that might mean defaulting to slowing or stopping when uncertain, rather than pushing limits. For a social media algorithm, this might mean not amplifying content until it's been verified as benign or giving users more control.

Emergent properties—outcomes that are more than the sum of their parts—are a central challenge of this frontier. We therefore need an ethical mindset of continuous learning and adjustment. Deploying an AI or network isn't a one-and-done; it's more like raising a child—continuous guidance and course correction as it grows and interacts with the world. Engineers and organizations need to commit to ongoing ethical reviews of their technologies post-deployment, much like companies do periodic security reviews. This could even take the form of an "ethics maintenance" schedule akin to equipment maintenance.

In the chapters ahead, we will delve into examples of operationalizing ethics. For now, from the exploration of AI, autonomy, and complexity, remember these key points:

- **We must design with accountability:** ensure that when machines make decisions, humans can trace and, if needed, challenge those decisions. This might involve logging decision paths, exposing options for appeal, and clearly defining who is responsible for each outcome.

- **We must insist on inclusive perspectives:** those who bear the risks of a piece of technology (often everyday people or marginalized groups) should have a say early on. This might be through user studies, public consultations, or having diverse teams that anticipate different impacts. As the saying goes, "Nothing about us without us" – don't impose an AI on a community without involving that community in the process.

- **Ethical trade-offs require debate:** There often isn't a single correct answer to questions like how safe is "safe enough" for an AV, or how much privacy to trade for pandemic safety. These are social decisions as much as technical ones. Engineers should be prepared to engage with policymakers and the public, explaining capabilities and limits candidly, so that informed decisions can be made. Ethical leadership sometimes means speaking up to regulators. For example, AI researchers might advise that certain uses of facial recognition by the government be curtailed because the risks outweigh the benefits.

- **Continuous improvement:** Just as we issue software patches, we should be ready to issue "ethical patches." If a bias is discovered, fix it and update the system. If users find an AI's behavior creepy or harmful, iterate on it. Ethical design is an ongoing process.

To conclude, the new frontiers of technology are testing us in unprecedented ways, but they also offer the opportunity to do tremendous good—if we can steer them with wisdom. AI can help eliminate human bias if done right. Autonomous systems can save lives, and complex networks can solve global problems like climate coordination. Achieving those upsides while avoiding the pitfalls is the crux of engineering ethics in our era. It requires expanding our thinking beyond the technical silos, anticipating failure modes at scale, and always remembering that technology is not an end in itself—it must serve human values. It must be guided by human empathy and conscience. In the next chapter, we will look at how to translate these ethical considerations into practical tools and processes that engineers and organizations can apply in their daily decision-making, thereby bridging the gap between ethical awareness and ethical action.

Key Takeaways from Chapter 2:

- **Algorithmic Bias and Fairness:** AI systems can unintentionally perpetuate or amplify biases present in their training data. Ethical practice demands rigorous testing for bias, inclusion of diverse data, and corrective measures to ensure AI decisions are fair and just.

- **Transparency and Explainability:** Explainability enables those affected to understand and contest decisions, and it allows engineers to debug and improve systems. Without transparency, people feel powerless, and errors or injustices can go unchecked.

- **Autonomous Systems and Safety:** When we hand over control to machines (self-driving cars, automated pilots, etc.), we must do so very carefully. Lives are at stake. Clear assignment of responsibility is needed for when things go wrong—we cannot accept a vacuum of accountability.

- **Systemic Complexity and Cascading Risks:** In tightly coupled systems, small failures can chain-react into catastrophes. Ethical engineering involves thinking beyond one's subsystem to the whole network of interactions. This means collaborating across disciplines and organizations to anticipate failure modes and protect the greater public. Building resilient systems (with redundancies, safeguards, and rapid response mechanisms) is not just good practice but a moral duty when so many depend on these networks.

- **Continuous Ethical Adaptation:** The fast pace of technological change means our ethical frameworks and regulations must evolve. Engineers and technologists should embrace a mindset of continuous learning and ethical improvement—monitoring how their innovations perform in the real world and being ready to adapt. Engaging with ethicists, policymakers, and the public is key to shaping technologies that reflect societal values.

Chapter 3:
Trust – The Invisible Grid of the New Terrain

Introduction

I have seen cities go dark and markets freeze. In those moments of crisis, the true measure of strength was simple: Could people still rely on one another? Could they stand steady when fear pressed in? When a massive regional power blackout struck, the electrical grid failed instantly—yet neighbors stepped up to direct traffic and share resources. When a financial panic hit and credit markets ground to a halt, algorithms and balance sheets were useless – only emergency action and public faith in institutions kept the economy from collapsing. The systems that failed were technical. The systems that held were human. Each time, it became clear that our most vital infrastructure is not steel or software; it is trust. Trust is the invisible currency that powers our grids, markets, democracies, and families. When trust collapses, no technology or wealth can save a society. This chapter explores how trust is built and broken, why it has become the fragile thread holding our world together, and how we can fortify this hidden grid before it frays beyond repair. This is another key reason why a commitment to ethics—cultivating resilience and building trust—matters more than ever.

We live in an age when trust is under strain everywhere. Global surveys paint a stark picture: only about 40% of people worldwide expect their families to be better off in five years—a steep drop in optimism compared to a few years ago.[3] Economic anxiety, disinformation, and polarization are eroding the glue that holds communities together. Public trust in governments and institutions is at historic lows in many countries. For example, in the United States, only around 20% of citizens now trust the federal government to do what is right—down from 77% in the 1960s, reflecting decades of deepening cynicism.[4] A worldwide "trust recession" is unfolding, driven by leadership failures, inequality, and rapid technological change. The World Economic Forum warns that the erosion of social cohesion and societal polarization—essentially a collapse of trust—ranks among the top global risks for the coming decade.[5] In an unraveling world of climate threats, cyberattacks, and geopolitical rifts, trust is both more precious and more precarious than ever.

Yet amid the uncertainty, there is hope: trust can be rebuilt. Throughout history and across societies, trust has proven to be a renewable resource—if we choose to renew it. Communities that nurture trust and social bonds have shown remarkable resilience in

---

[3] Joe Myers, "Trust barometer: 4 steps to rebuilding trust," World Economic Forum, January 18 2023, https://www.weforum.org/stories/2023/01/edelman-trust-barometer-rebuilding-trust/.

[4] Joe Davidson, "American trust in government near 'historic lows,' Pew finds," *The Washington Post*, June 9, 2022, https://www.washingtonpost.com/politics/2022/06/09/american-trust-government-pew-survey/.

[5] Myers, "Trust barometer."

the face of disasters and turmoil. Leaders who earn trust through integrity and empathy have guided nations through dark hours. In the chapters ahead, we will examine trust as the cornerstone of effective systems (from power grids to supply chains), the key to ethical leadership and social stability, the linchpin of resilient communities, and the unwritten code passed across generations. Through storytelling, personal experiences, and analysis of real cases on every continent, we will see why our future depends on what we cannot see—the currency of trust—and how investing in trust is a key part of being an ethical leader today.

The Currency of Trust: The Invisible Fuel of Systems

Trust is not just a virtue—it is the currency that powers every system we rely on. In the modern world, nearly everything runs on trust. We take for granted that flipping a light switch will bring electricity, that money in the bank will be available when needed, or that a posted letter will reach its destination. Beneath each of these expectations lies a bedrock of trust in people, institutions, and processes which we cannot personally verify every time. Economist Kenneth Arrow once observed that "virtually every commercial transaction has within itself an element of trust."[6] Indeed, trust is the invisible fuel that keeps commerce, technology, and society functioning. When it falters, systems grind to a halt—sometimes overnight.

Consider the global financial system, which essentially runs on trust. Money itself is a promise, valuable only because people believe in its value. During the 2008 global economic crisis, that

---

6   Esteban Ortiz-Ospina, Max Roser, and Pablo Arriagada, "Trust," Our World in Data, April 2024, https://ourworldindata.org/trust.

belief was shaken to the core. Banks stopped trusting one another's solvency, and the interbank lending that greases the economy froze up. As one analysis noted at the time, "Banks don't fully trust each other… there's just an environment of distrust right now, and that's the core of this entire crisis."[7] Interest rates for overnight loans between banks spiked to record levels as each institution feared the next might collapse. Credit—whose Latin root *credere* means "to believe"—evaporated. It took massive government interventions to halt the downward spiral. The lesson was stark: when trust vanishes, so does liquidity; markets seize up, and even enormous wealth cannot be substituted for the missing trust. What holds a complex financial network together is the shared belief that obligations will be honored. Once that faith is lost, the network unravels. The 2008 meltdown proved that elaborate financial engineering is no match for a fundamental loss of confidence. In the end, trust *is* the capital that matters most.

The same is true for other critical systems. Energy grids, supply chains, and communications networks all require layers of trust to function smoothly. Our electric power grid is often called the most complex machine ever built—made up of thousands of power plants and millions of miles of lines working in unison. But the real glue is the trust between all actors: that grid operators will keep supply and demand balanced, that maintenance crews will do their jobs, that customers will pay their bills, that regulators will enforce safety. When that trust is absent or broken, even the best technology can falter. For instance, during widespread power outages, what often determines whether society stays calm or descends into chaos

---

7   Neal Irwin, "Mutual distrust freezes lending among banks," *The Washington Post*, September 30, 2008.

is public trust. In one stark comparison, New York City's blackout of 1965 (as well as a later outage in 2003) was met with relative calm and community cooperation, while the blackout of 1977—coming amid fiscal crisis and frayed social trust—resulted in an eruption of looting and arson across dozens of neighborhoods.[8] The technical cause in each case was similar (a grid failure), but the social aftermath diverged dramatically. Trust was the variable: in 1977, many New Yorkers had lost faith in city leadership and economic justice, and when the lights went out, civic order collapsed. In 1965 and 2003, by contrast, a stronger residue of trust and social capital helped people pull together despite the darkness. These episodes reveal that the true resilience of infrastructure lies in human relationships. A power grid can fail safely if people trust each other —or become a catastrophe if they do not.

Look at any efficient supply chain, and you will see trust holding it together. Whether it's a farmer trusting a distributor to pay on delivery, or a factory trusting its parts supplier to meet quality standards, each link relies on faith in the next. The COVID-19 pandemic provided a dramatic illustration: at the outset, many nations and states engaged in frantic bidding wars for masks and medical supplies, often out of distrust that others would share fairly. Global supply chains were strained by protectionism and suspicion. In contrast, regions that coordinated and trusted each other's data—for example, scientists sharing vaccine research openly—achieved faster breakthroughs. In normal times, too, businesses depend on reliable partners and a stable rule of law to enforce contracts. When trust in those foundations erodes, companies either withdraw or

---

8 "New York City blackout of 1977," Wikipedia. Accessed August 18, 2025. https:///wiki/New_York_City_blackout_of_1977

incur steep costs protecting themselves. Economists have found a strong correlation between societal trust levels and economic prosperity.[9] High-trust societies tend to have lower transaction costs, smoother business cooperation, and more investment. Low-trust environments, on the other hand, must expend enormous effort on verification, security, and litigation, which acts like sand in the gears of commerce. In this way, trust is a form of capital—social capital—that boosts efficiency and growth. A handshake deal in a high-trust community can be more valuable than a legal contract in a low-trust one.

Zooming in further, even our families and personal lives run on trust as currency. From the time we are children, we learn to trust (or distrust) the world around us, and those early experiences form our "human code" for life. A child who grows up in a trusting, supportive environment gains confidence to explore and collaborate, whereas a child raised amid broken promises and betrayal may carry deep skepticism. In workplaces, teams with high trust among members consistently outperform those mired in office politics and fear. Marriages and friendships thrive on trust and crumble when it is gone. In communities, neighbors who trust each other can create informal safety nets—watching each other's homes, helping with childcare —that no government program can replicate. During disasters, these trust-fueled networks become literal lifelines. Sociologists have documented how communities with strong interpersonal trust and civic engagement cope far better with crises. People share resources, evacuate the vulnerable, and rebuild together. In contrast, communities divided by suspicion may see "everyone for

---

9   Daniel P. Aldrich, Building Resilience: Social Capital in Post-Disaster Recovery (University of Chicago Press, 2012).

themselves" behavior that makes the crisis far worse. Trust, at its heart, is a mutual belief in reliability and goodwill. It allows us to overcome the inherent uncertainty in any interaction. Without that belief, cooperation collapses to its lowest denominator—contracts and coercion—which are costly and brittle substitutes.

In short, trust is the unseen currency spent in every human exchange. It is the confidence that others—known or unknown—will do the right thing. It greases the everyday machinery of life, from buying groceries with paper money to electing leaders to managing a public health crisis. We notice trust only when it's missing: when a scam makes us wary of strangers, when a failed promise leaves us jaded, or when a breach of integrity makes headlines. Only then do we realize how much of our world runs on assumptions of good faith. As we will explore, those assumptions are being tested today as never before. But first, we must understand how fragile this currency can be, and what happens to societies when the bank of trust runs dry.

The Fragile Thread: Trust and Survival in an Unraveling World

History shows that civilizations do not usually fall from a lack of power or wealth – they fall when the thread of trust binding them breaks. Great powers have been defeated by internal rot long before external enemies overran them. Ancient Rome had formidable legions and vast treasure, yet a decay of civic virtue and trust marked its decline. As corruption spread and the social contract frayed, ordinary Romans lost faith in their leaders and the unity of the empire. By the end, many communities did not bother to defend Rome against barbarian invaders—the allegiance to Rome had already died in their hearts. In the same vein, the Soviet Union did not collapse in 1991 because it ran out of tanks or missiles. It collapsed when its people ceased to believe in the system holding it together. A popular dark joke in the USSR at the time captures this loss of trust: "They pretended to pay us, and we pretended to work." Decades of propaganda, economic stagnation, and broken promises had eroded any trust the Soviet citizens had in their government. When reformist leaders tried to introduce openness (glasnost) and restructuring (perestroika), it was too late—the foundation of trust was gone, and the union fell apart with shocking speed. These examples underline a vital point: when trust dies, systems of any size—empires, governments, companies—become meaningless. They may linger on for a time, but they are one shock away from collapse because no one is truly invested in their success.

Today, that fragile thread of trust is under strain across the world. We see the warning signs in our headlines and our communities. In many democracies, public trust in elected leaders and institutions has hit record lows, fueling a populist wave of anger. People who feel lied to or left behind by elites become susceptible to demagogues who promise easy fixes while further poisoning the well of trust. For instance, surveys in country after country show deep skepticism that governments are working in the people's interest—a view reinforced by high-profile scandals of corruption and abuse of power.[9] Each scandal is not just a single breach; it accumulates into a narrative that "the system is rigged" or "politicians cannot be trusted," driving citizens into cynicism or even onto the streets in protest. In authoritarian states, the trust problem takes a different form—an enforced facade of trust maintained by fear. Because dissent is punished, people may profess loyalty but privately doubt the regime. Such top-down "trust" is brittle: when cracks appear, they can quickly widen into collapse, as seen in Eastern Europe's revolutions. Genuine trust cannot be commanded; it must be earned.

The fragility of trust is evident even in seemingly stable systems. Modern infrastructure often gives an illusion of invincibility—until a crisis hits. When natural disasters strike—a massive hurricane, a regional blackout, a pandemic—they reveal the underlying condition of social trust. Communities with strong trust and social networks tend to respond effectively: neighbors check on neighbors, volunteers mobilize, and there is faith that help will come. In such communities, people often self-organize to fill gaps, demonstrating resilience. Research has shown that social capital (the networks and trust within a community) is a top predictor of disaster recovery success. After Japan's 2011 earthquake and tsunami, for example, villages with higher levels of civic engagement and trust had significantly lower mortality rates and recovered faster than those with weaker social ties.[10] In contrast, where trust is low, crises can spiral into social breakdown. We have witnessed instances where rumors incite panic buying, or residents refuse to evacuate because they don't believe officials. During Hurricane Katrina in 2005, the slow and chaotic government response, combined with pre-existing distrust, especially among poorer communities in New Orleans, led to a collapse of order in parts of the city. Misinformation and distrust hampered coordination, and it took years for trust in institutions even partially to mend after that failure. Simply put,

---

10 Aldrich, Building Resilience.

resilience is not just about robust physical systems, but about robust social trust. A community might have strong levees or stockpiles, but if people don't trust the authorities or each other, those assets can be squandered or underused.

On the global stage, the absence of trust between nations can mean the difference between peace and conflict. Treaties and alliances are only as strong as the credibility of the parties involved. The post-World War II international order—institutions like the United Nations, NATO, and the European Union—was built on the hard-earned trust that cooperation was better than war. But in recent years, cracks have appeared. Longstanding allies question each other's commitments (as seen when trust wavered in transatlantic relations), and rivals exploit disinformation to sow distrust. The risk is that without a basic thread of trust, even imperfect, the world could slide into a new era of fragmentation and rivalry, undermining progress on issues that require unity (from pandemics to terrorism to climate). Indeed, the theme of the World Economic Forum's 2023 summit was "Cooperation in a Fragmented World," reflecting the urgent need to rebuild trust and cooperation in the face of fragmentation.[11]

The Hidden Grid of the 21st Century

Beneath all the visible grids and networks of our modern age lies an invisible one: the grid of trust. We often think of infrastructure in terms of concrete, wires, code, and policies. But an electrical grid or an information network is ultimately a framework that depends on people doing their part reliably. Trust is the unseen circuitry that links all the components together. Although it doesn't appear in any

---
11 Myers, "Trust barometer."

schematic diagram, its failure can cause the whole system to crash. The Twenty-first century has brought astounding technological connectivity—we can transact, communicate, and travel across the globe faster than ever. Ironically, this hyper-connectivity makes the trust grid even more critical. With so many interdependencies, a single failure can cascade widely. The only way complex systems don't "break all at once" (as one grid expert said in a post-9/11 briefing) is if there are buffers—and trust provides many of those buffers.[12]

Think of trust as the bandwidth in a societal system. High trust means information and cooperation flow quickly, like data on a high-speed connection. Low trust is like narrow bandwidth—everything slows down with additional verification, oversight, and friction. Trust is also akin to voltage stability in the power grid: it keeps the current of civic life flowing steadily, preventing small disturbances from turning into blackouts. When trust is high, a society can absorb shocks—people give each other the benefit of the doubt and pull together in emergencies. But if trust is already low, even a minor shock (a rumor, a temporary outage, an election result) can trigger disproportionate turmoil. In that sense, trust is the safety margin that prevents systemic failures. A society with no trust margin is always one incident away from crisis.

Moreover, trust acts as the operating system for collaboration. Just as computers need a common OS to coordinate processes, humans need trust to coordinate at scale. I recall working on community resilience workshops in a city where we mapped "heat zones" of trust: we asked residents, "Who would you turn to if the system fails?" The answers (the neighbor with a generator, the

---

12 Aldrich, "Building Resilience."

corner store owner who extends credit, the old church that still has a landline) revealed an invisible web of relationships that was the true first-responder system. When we showed city officials these community trust maps, it was a revelation. The city's formal plans had overlooked the informal trust grid that holds things together when formal systems falter. As one community organizer succinctly put it, "Relationships are the real infrastructure."

Because trust is largely invisible, it can be woefully undervalued by decision-makers focused on tangibles. A company might slash employee benefits to cut costs, not realizing it is chipping away at workers' trust and thus long-term productivity. A government might roll out a policy without community input, saving time upfront but breeding mistrust that will cost dearly in enforcement and compliance problems later. These are examples of "efficiency" that backfire by weakening the hidden grid. Truly forward-thinking leadership, in contrast, treats trust as a critical asset to invest in. For instance, some police departments are shifting from pure enforcement metrics to "trust metrics"—measuring community sentiment and engagement—on the understanding that public safety ultimately flows from public trust. In business, the rise of corporate ethics and sustainability efforts reflects an understanding that trust from consumers and society is a competitive advantage, driving everything from brand loyalty to the ability to attract talent.

Investing in trust can take many forms. It might mean corporations embracing transparency —sharing data on their supply chains or environmental impact—even if not legally required, to earn public confidence. It can mean governments practicing radical honesty with citizens about challenges and mistakes, instead of spinning or covering up, thereby building credibility over time. It certainly means leaders being willing to stand in truth even when it costs them, because nothing builds trust like visible integrity, and nothing destroys it faster than deceit.

In this age of disruption, we have seen how brittle the old definitions of strength are. Military might, economic size, and technological prowess—none guarantee resilience if trust is absent. The mightiest power grid fails if people don't cooperate to conserve energy in a crisis. The strongest economy falters if consumers and banks panic and withdraw trust. The most advanced democracy can teeter if citizens refuse to trust election results or each other's basic decency. The future, it turns out, will not be won by the society with the biggest arsenal or the fastest supercomputers. It will be won by those who can still be trusted when everything else breaks. Trust is the factor that turns disparate individuals into a collective capable of overcoming challenges. It is what allows us to take that leap of faith to cooperate, innovate, and sacrifice when needed.

The hidden grid of trust carries more weight than concrete, steel, or code. Every decision we make, every interaction, is either building it up or wearing it down. The choices might be as small as being honest in a business deal or as large as a nation owning up to a historical injustice —but each one matters. If we treat trust as our most precious infrastructure and nurture it with the same care that we give to our roads and bridges, we will create systems that endure. If we squander it, we risk a future of constant breakdowns and fragmentation.

Conclusion: Rebuilding Trust, Rebuilding the Future

In the end, trust is the ground beneath every step we take as a civilization. It is the silent pact that turns chaos into cooperation and adversity into collective action. Break that ground, and nothing lasting can be built. Keep it strong, and even in ruin, a path to recovery can be found. The currency of trust is unique: the more we spend it in good faith, the richer we become. Unlike wealth or technology, trust is not zero-sum—it multiplies with use, forging resilience in families, teams, and nations.

Looking ahead, we face a fork in the road. Down one path, societies continue to splinter as trust in each other and in institutions crumbles. In this scenario, problems multiply: democracies gridlock or backslide into authoritarianism because social trust—the willingness to compromise and see others as legitimate—disappears. Economies stagnate as investors and consumers retreat out of fear. Communities balkanize, each withdrawing behind gates or into online echo chambers. Instead of connecting us, technology becomes a battleground of misinformation and surveillance, further eroding trust. In the worst case, the loss of confidence could even lead to violence and conflict, as groups demonize each other and no longer believe in peaceful solutions. This grim future is not a sci-fi dystopia; elements of it are already visible wherever trust has collapsed.

But down the other path, we recognize the value of trust and act decisively to restore it. This path is harder but far more hopeful. It involves leaders being accountable and transparent, demonstrating ethics that earn back public confidence. It affects communities actively investing in social bonds—fostering dialogue across divides, building inclusive institutions, and remembering that trust is built slowly, through consistent honesty and empathy. It involves redesigning systems—from policing to social media algorithms—with trust and fairness in mind, so they don't inadvertently breed cynicism or inequality. Perhaps most importantly, it involves each of us making daily choices that strengthen trust: keeping our word, listening to others' concerns, and showing up for our neighbors. Trust, after all, is contagious. Small acts of trustworthiness can ripple outward, inspiring others to do the same.

If there is one message to carry forward, it is a hopeful one: trust can be rebuilt. We know this because it has been done before. After eras of conflict, nations have reconciled and built trust anew—as Germany and France did after 1945, laying the foundation for a peaceful Europe. After periods of corruption and abuse, institutions have been reformed and public trust earned back—as seen in various anti-corruption successes from Singapore to Georgia, where determined leadership broke the cycle of cynicism. Communities torn by disaster or violence have found healing through truth and reconciliation, whether through South Africa's Truth and Reconciliation Commission or grassroots peace circles in war-torn villages. These stories show that even when trust has been gravely damaged, it is not beyond repair. It requires truth-telling, courage, and time, but the thread can be reforged.

For general readers, policy makers, business leaders, and young professionals alike, the implications are clear. Trust is not a "soft" sentiment to consider after the real work is done—it *is* the real work. A policy that ignores trust, an innovation introduced without trust, or a strategy that undercuts trust is ultimately self-defeating. Conversely, an initiative that builds trust can transform what is possible. This applies whether you are leading a multinational corporation, governing a city, or simply trying to support your neighbors. The specific tools will differ—it might be adopting a community-driven planning process for a new project, implementing ethical guidelines for AI, practicing radical candor with your constituents, or mentoring youth in your neighborhood—but the end goal is the same: to weave a stronger fabric of trust.

Our future truly depends on what we cannot see. It depends on trust—the currency that carries our hopes and the thread that ties our fate. If we value it, protect it, and grow it, there is no storm we cannot weather, no grid we cannot keep alive. As you navigate the emerging ethical terrain, trust will be the line that holds. Let the next section serve as your map to guide you on your journey. May you be led by a commitment to follow the through-line of trust. May you never let it go.

Key Takeaways from Chapter 3

- **The Deeper Infrastructure:** Trust is the invisible currency that powers our lives on a societal, familial, and personal level. When trust collapses, no technology or wealth can save the system. The true infrastructure is relationships.

- **The Consequences of Lack of Trust:** Whether in a business transaction or a government initiative, high trust means information and cooperation flow quickly. With low trust, everything slows down with additional verification needed at each step.

- **Build-in Trust:** Truly forward-thinking leaders, whether in a business or other setting, treat trust as a primary asset to invest in.

- **Learning from History:** While levels of trust globally are at all-time low, trust is a resource that can be rebuilt (as past historical examples show).

Part II:
The Map

Chapter 4:
Embracing Failure – The Inventor's Path

A Mirror & a Map: The Power of Stories To Guide Us

I still remember the first time I stood in front of a classroom full of bright-eyed students, armed with a stack of textbooks and theories. I was supposed to teach about management and leadership, innovation, and change. The theories were sound, and the case studies were neatly summarized in bullet points. Yet, as I began to speak, I could sense a gap between the polished principles on the page and the messy, vibrant reality of the world outside. Over the years, through my teachings and presentations, I discovered a simple truth: nothing conveys a lesson more powerfully than a real story told with authenticity. Theory sparks interest, but stories spark change.

We as humans are hardwired to respond to narratives. When we hear about an inventor persisting through failure or a community transformed by an inventor's vision, something clicks. The lesson stops being abstract and becomes tangible and memorable. I noticed that when I shared real cases involving real people and institutions, my students leaned forward. They asked questions. They reflected on what they would have done in that situation. In those moments, learning became actionable-something they could imagine applying in their own lives or careers. It's as if the story provides a mirror and a map: a mirror that reflects challenges we all face, and a map that hints at how to navigate those challenges.

In this section, I have collected some of the most compelling stories I've encountered or used in my teachings. These stories span different industries, cultures, and contexts-from corporate boardrooms to rural villages-yet they all carry universal lessons. You will read about innovators who turned failure into success, leaders who upheld trust and ethics in crisis, companies that learned to adapt or perished because they did not, and communities lifted up by empowerment and technology. Each chapter dives into a specific story in rich detail, letting the voices of the people involved come through. You'll hear from visionary founders, skeptical workers, determined community members, and others-in their own words when possible-because their perspectives bring each lesson to life. Each serves as an example of ethics in action, whether by cultivating resilience or tapping into trust.

I've also included reflections and analysis after each story, to draw out key insights and actionable lessons you can carry forward. The goal is not to preach one single "right way" to lead or innovate, far from it. Instead, the goal is to invite you to step into these real scenarios, reflect on them, and extract wisdom that is relevant to your journey.

A few themes echo throughout the chapters. One is authenticity: the notion that honest reflection on failures and successes provides the richest learning. Another is actionability: after each story, we won't just admire what happened; we will pinpoint how you can apply similar principles in your context. You will find bullet-point Key Takeaways at the end of each chapter for a concise summary of lessons and practical advice. This should help translate inspiration into action, whether that means changing your leadership style, approach to problem-solving, or outlook on what's possible.

So, whether you are a student, a professional, a leader of a team or community, or simply someone hungry for insight, I invite you on this journey. As you read, imagine yourself inside each story. Feel the frustrations, the hopes, the doubts, and the triumphs of the people you'll meet in these pages. Let their voices speak to you. And as you turn the last page of each chapter, take a moment to ask: What would I have done? What can I do now in my sphere of influence with this lesson? If this book spurs you to take one concrete action-embrace a failure, listen more closely to others, take a principled stand, try a new approach-then our shared journey of learning will have been worthwhile. I offer these stories as guideposts, examples of how others have navigated the ethical terrain in their own way, to serve as an entry point for your own journey.

With that, let's step into the first story, which begins not in a boardroom or a classroom, but in a humble workshop filled with the whir of machines and the crumpled sketches of an inventor who refused to give up and cultivated resiliency.

The Inventor's Path to Success

"I made 5,127 prototypes of my vacuum before I got it right. There were 5,126 failures. But I learned from each one. So I don't mind failure." These words were spoken by Sir James Dyson, the British inventor and entrepreneur, reflecting on the arduous journey that led him to invent the world's first bagless cyclonic vacuum cleaner. Today, Dyson's name is synonymous with innovation-his products range from high-end vacuum cleaners to hand dryers and hair stylers-but his success didn't happen overnight. It was built atop a mountain of failures. If you walked into Dyson's workshop in the late 1970s and early 1980s, you would have seen a tinkerer's paradise cluttered with metal parts, plastic molds, and dozens of prototypes of strange-looking vacuum devices. Many of those prototypes were nonfunctional; some fell apart in testing. Yet each one taught Dyson something new. This chapter explores the story of James Dyson and how embracing failure as a learning process can be a powerful driver of innovation and eventual success.

The Long Road of Trial and Error

James Dyson was not initially a vacuum magnate or a household name. In the early 1970s, he was a young engineer who became frustrated with how traditional vacuum cleaners lost suction as their bags filled with dust. Determined to solve this problem, Dyson took inspiration from an unlikely source: industrial cyclones used in sawmills to separate sawdust from air. He envisioned a vacuum cleaner that could use centrifugal force to spin dirt out of the air without needing a porous bag that would clog. It was a radical idea at the time.

Armed with this concept, Dyson built his prototype. It was crude, made of cardboard and tape, but it showed a glimmer of promise. The dirt spun out just as he thought it might. However, the design was far from practical—dust blew everywhere, and the contraption was bulky. Dyson didn't see these shortcomings as signs to quit. Rather, he saw them as feedback. Each failure highlighted a flaw that needed fixing. Instead of giving up, he refined the design again. And again. And again. He worked in a garage behind his house, often late into the night, fueled by a stubborn belief that he was onto something revolutionary even as people around him questioned his sanity. Family and friends saw him sweating over one model after another, accumulating debt, and chasing what looked like a futile dream. But Dyson persisted. By his count, he built 5,127 versions of his vacuum over five years. Five years of failures—or as Dyson viewed it, five years of education.

One can only imagine the emotional ups and downs in that process. In interviews, Dyson later shared how he had experienced moments of deep frustration, but strangely, also of enjoyment: "I enjoyed those failures," he said. "Every time something broke or didn't work, I thought, 'Aha, I've learned what doesn't work. I'm a bit closer to finding what will." This mindset of treating failure as a teacher rather than an enemy became the cornerstone of Dyson's approach. It's a trait shared by many great inventors and innovators through history. Thomas Edison, when working on the light bulb, famously said, "I have not failed. I've just found 10,000 ways that won't work." Dyson was cut from the same cloth. His perseverance was not blind stubbornness-he iterated with purpose. Each of the 5,127 prototypes introduced a small change: a tweak in the cyclone cone angle, a different material for the filter, a new way to collect the dust. Most of those changes didn't produce the breakthrough he needed. But little by little, they collectively paved the way for success.

Breakthrough and Rejection

By prototype number 5,127, James Dyson finally had a working design that met his goal. The device maintained strong suction using dual cyclones and had a clear bin to collect dust (so you could see when it needed emptying). It was clever and effective. Dyson was ecstatic-he had done it! He had turned persistence into a tangible innovation.

However, the story doesn't end with the invention. Another form of trial awaited him: market rejection. When Dyson approached established vacuum cleaner companies to license his bagless design, he was turned away repeatedly. Manufacturers did not welcome an innovation that could disrupt the lucrative market in which they had been selling replacement vacuum bags. They likely thought, "Why introduce a product that doesn't require bags, when bags are our ongoing revenue stream?" This short-sighted thinking led to Dyson hearing "no" time and again from the very companies that could have helped bring his invention to consumers. It was another series of failures-this time not of a prototype, but of persuasion. Many people would have become disheartened at this point. Despite all those years and prototypes, the industry's doors remained closed.

Dyson, true to form, did not give up. If nobody would license his invention, he decided he would produce it himself. He took a big financial gamble, eventually securing loans and a mortgage on his house to start manufacturing the vacuum, which he called the "Dyson Dual Cyclone." In 1993, he launched it in the UK under his own fledgling company. Early marketing emphasized that the Dyson vacuum never lost suction and required no bags. At first, retailers were skeptical—this was a vacuum cleaner that cost significantly more than the average, made by a brand no one had heard of, and using technology no one had seen. But word-of-mouth testimony from satisfied customers grew, and reviews were positive. Within a few years, the Dyson vacuum became a status symbol for cleanliness and innovation. By the late 1990s, Dyson's product overtook well-known brands to become the top-selling vacuum in the UK. Eventually, Dyson entered markets worldwide, including the United States, to great success. The man who had once been labeled a "mad inventor" laboring in his coach house became a billionaire entrepreneur and was knighted for his contributions to engineering.

It's easy now to see only the result and forget the sweat and failures. But Dyson himself never forgot. Even as his company expanded into lighting, fans, hairdryers, and more, he instilled a culture in his company that does not fear failure. Engineers at Dyson reportedly create hundreds of prototypes for each new product, testing and discarding many ideas to find the one that works best. The lesson Dyson carried forward is that every failure is an opportunity in disguise.

The Gift of a "Failure-Friendly" Mindset

Dyson's story illustrates the value of perseverance, but more deeply, it highlights what we might call a "failure-friendly mindset." In a traditional school upbringing (as Dyson has often noted), we are taught to avoid failure, to get things right on the first try. But real life is different. Complex problems rarely have clear, immediate solutions. In Dyson's words, "It's a life of failure. It's trial and error.

When something finally works, it's less interesting because you've solved it. The interesting bit was the failures!" This counterintuitive viewpoint suggests that if we can learn to see failure not as a verdict on our abilities but as part of the process, we free ourselves to innovate without paralyzing fear.

A similar mindset can be found behind the creation story of other products as well. Consider the popular lubricant and water-displacer spray WD-40. The quirky name "WD-40" actually stands for "Water Displacement, 40th formula." Norm Larsen, the chemist who invented it in 1953, failed on his first thirty-nine attempts to create a formula that would prevent aerospace parts from rusting by displacing water. He succeeded on the fortieth. Instead of hiding those thirty-nine failures, the product's very name celebrates them. It says to everyone who uses it: if at first you don't succeed, try, try again (and maybe try a few dozen times more!).

Or think of the Post-it Note, a ubiquitous office product born from a mistake: a 3M scientist set out to create a super-strong adhesive but accidentally made a very weak, reusable one. For a while, that "failed" adhesive had no obvious use until another colleague thought of anchoring notepapers with it, and voila—Post-it Notes were born. In each of these cases, a supposed failure was not the end of the story, but the beginning of a new one.

What makes someone like Dyson or Larsen persist through hundreds of failures while others give up? Part of it is passion—they deeply care about the problem they are trying to solve. Part of it is confidence, not necessarily confidence that they know the answer (they often don't at first), but confidence that an answer exists and can be found if they keep at it. Another part is their relationship to risk and learning. They do not see a failed attempt as a personal humiliation or a waste of time, but as an integral investment in understanding and a key part of building resilience

James Dyson also benefited from a certain degree of freedom to fail. Early in his career, he took a big risk and started on his own, which meant he didn't have a boss scolding him for each failure. Unfortunately, in many corporate or academic environments, failure is stigmatized. But Dyson's story invites us to rethink that. What if organizations rewarded well-intentioned failures? What if managers asked after a project flop, "What did we learn that moves us forward?" rather than only, "Whose fault was this?" A "failure-friendly" culture can unleash creativity because team members stop playing it safe and start experimenting, which is part of being able to respond adeptly to the unforeseen—and to change.

Learning to Fail Intelligently

Embracing failure doesn't mean being sloppy or repeating the same mistakes without alteration. Dyson's prototyping process was methodical. He practiced intelligent failure, which involves forming hypotheses, testing them, learning, and adjusting. Each prototype was a controlled experiment. If he changed five variables at once and it worked, he wouldn't know which change mattered. So often he changed one variable at a time to isolate its effect. This way, even a complete dud of a prototype yielded insight. Similarly, in any field, one can plan small experiments or pilot projects that probe an idea's validity. If they fail, it's on a small scale (low cost, low damage) but still yields information to guide the next step. This concept is prevalent in modern approaches like agile software development (where quick iterations and "fail fast, learn fast" are encouraged) and design thinking (which promotes early prototyping and testing with users to incorporate feedback).

Dyson's saga also shows the importance of resilience-the capacity to bounce back from setbacks. There must have been mornings he woke up after the hundredth or thousandth prototype failed, feeling defeated. Resilience provided the push to start number 101 and number 1001. It is notable that Dyson launched and later abandoned other projects as well. For instance, in 2017, he began developing an electric vehicle under the Dyson brand, investing significant resources. A couple of years in, he concluded that the project was not commercially viable and made the tough decision to cancel it. "The route to success is never linear," Dyson wrote to his employees, explaining the move. "This is not the first project that has changed direction, and it will not be the last." In that decision, we see that embracing failure also means knowing when to cut losses. Stubborn persistence must be balanced with the wisdom to pivot when evidence mounts that an idea won't pan out. Dyson didn't view the electric car venture as wasted effort; rather, it was an exploration that revealed new knowledge (and possibly spurred some battery technology advances they can use elsewhere).

Voices of others who have trodden similar paths echo Dyson's mantra. Silicon Valley entrepreneurs often speak of celebrating failure. A failed startup, they say, is not a badge of shame but a rite of passage on the road to eventual success, provided you learn from it. There are even events called FailCon (Failure Conferences) where founders share candid stories of their failures to help others avoid the same pitfalls. The idea is not to glorify failing for its own sake, but to remove the stigma so that innovators are not afraid to take risks. Taking risks in turn is part of strengthening the ability to not only accept change but to thrive on it.

Key Takeaways from Chapter 4:

- **View Failure as Feedback:** Every failed attempt is rich with information. Shift your mindset from "I failed" to "I discovered one way that doesn't work." This keeps you moving forward and refining your approach.

- **Practice Intelligent Iteration:** Like Dyson's prototype process, change one thing at a time and observe results. Structured, incremental experimentation makes failures manageable and educational rather than catastrophic.

- **Cultivate Resilience:** Innovators need grit. There will be moments of doubt and external skepticism. Don't forget why you started and push through setbacks, adjusting your path but not your overall vision.

- **Create a Safe Space for Trial and Error:** Whether you're an individual or leading a team, allow room for mistakes. Encourage the sharing of lessons learned from failures. A no-blame, learning-focused culture leads to more innovation.

- **Know When to Pivot:** Embracing failure doesn't mean mindlessly repeating the same failing strategy. Be prepared to pivot or halt a project when evidence and learning indicate that a fundamentally different approach is necessary.

By understanding that failure is not the opposite of success but a part of its journey, we liberate ourselves to be truly creative and persistent. James Dyson's story is living proof that tenacity and learning from failure can eventually turn a dust-filled idea in a garage into a world-changing product. In the next chapter, we will explore another side of learning—one that comes not from our trials and errors, but from listening to the voices of others who are affected by our decisions.

Chapter 5:
Listening to Stakeholders – A Water Pump Lesson

In the early 2000s, an ingenious idea captured the imagination of philanthropists, politicians, and the media. It was called the PlayPump: a merry-go-round attached to a water pump, designed to harness the energy of children's play to pump clean water for African villages. On paper, it sounded like a dream solution. Children get a playground toy, villages get a source of water with minimal effort, and donors get a feel-good story of innovation. Major figures rallied around the concept; First Lady Laura Bush championed it, big donors like Steve Case (founder of AOL) poured in millions, and celebrities such as Jay-Z visited PlayPump sites with TV cameras in tow. A whirlwind of positive publicity ensued. By 2006, a plan was in motion to install thousands of PlayPumps across sub-Saharan Africa. Yet, only a few years later, the dream crumbled. Many of these pumps fell into disuse, and some villages quietly removed the merry-go-rounds to reinstate their old hand pumps. What went wrong? The case of the PlayPump is a powerful lesson in the importance of listening to stakeholders, especially the voices of those on the ground whom an innovation is supposed to serve. It reminds us that even the best-intentioned, technically clever solutions can fail if they are designed for people, but not with people. A truly ethical vision includes stakeholders as part of considering the larger systemic impact (and as part of building trust).

The Promise of an Innovative Solution

Let's first picture the scene before the PlayPump was introduced. Imagine a rural community in Mozambique or South Africa, where traditionally women (and sometimes children) spend hours a day pumping water from a borehole well using a manual hand pump or even fetching water from distant streams. It's hard work; it's time-consuming. The PlayPump promised to change that narrative. In this new scenario, children at a school or in the village would push a brightly painted roundabout. As they spun around gleefully, their motion would drive a pump that sends water up into a storage tank. That tank, perched a few meters above ground, had a tap, so anyone could fill a container from it as needed. The storage tank even had four large billboard faces on its sides; the idea was to sell advertising (some of it for public health messages, some for commercial ads) and use that revenue for maintenance costs, making the system self-sustaining financially. It was a symphony of win-win scenarios: kids have fun, water is pumped for free, maintenance pays for itself, and donors get to save lives by installing a piece of joyful playground equipment. No wonder it won accolades. In 2005, a PBS Frontline documentary segment showcased the PlayPump in action, portraying it as a breakthrough for water access. The piece elicited enthusiasm worldwide.

Trevor Field, the South African advertising executive-turned-social entrepreneur behind PlayPump, became its tireless promoter. "If we could put 1,000 pumps in each water-stressed country," Field proclaimed, "we'd make a monster difference to rural water supplies." He had a vision of scaling this invention far and wide. Money followed vision. The U.S. government committed $10 million, Steve Case's foundation chipped in $5 million, and campaigns were aiming to raise over $60 million more. By late 2006, the PlayPump had gone from a local innovation to an international cause célèbre.

Voices from the Ground

Now, fast forward a few years. In 2009, journalist Amy Costello, who had reported the initial glowing story, returned to some PlayPump sites in Mozambique to see how things were going. What she found was sobering. At one school, the merry-go-round still stood, but "the children were standing idle," she noted. The storage tank was often empty. Why? "We don't know why, but no water is stored in the tank," an assistant principal told her. It used to work, but not anymore.

As Costello traveled to other villages, she encountered the people the fancy New York fundraisers and D.C. press conferences had overlooked: the local women and grandmothers drawing water. Their perspective was illuminating. One older woman, named Regina, pointed at the colorful roundabout with a mix of frustration and resignation. The PlayPump hadn't produced water in six months, she said. When it did work, it was hard to operate for an adult, far harder than the simple hand pump they used to have. "No one consulted us about this change," another woman explained. "The PlayPump just arrived." Those words are worth reflecting on: no one consulted us. In the rush to implement what sounded like a brilliant idea, the implementing organizations had failed to ask the very people who would rely on the pump what they needed, wanted, or could manage.

The women told the reporter that children didn't play on the pump as consistently as they had imagined. Why would they? To provide enough daily water for a community, kids would have to play for hours and hours on end. That might happen during a special event or if visitors (or cameras) are around, but it's not a daily occurrence. Children have school and chores, or they might simply get bored or tired. So, guess who ended up trying to spin the merry-go-round to get water? The same women who used to work the hand pumps. Except now, the mechanism was awkward for them. Pushing a heavy metal roundabout in circles is not the natural stance for drawing water, and especially for older women, it was difficult. Some would put their backs against it and walk to try to rotate it, essentially turning play into work, backbreaking work. "The old hand pumps were much easier," they said. In some places, when the PlayPump broke down (as all mechanical systems eventually do), it sat unusable for long stretches. One report later revealed that pumps were out of service for up to seventeen months waiting for repairs. With the old system, a villager could often fix the hand pump with basic tools, or spare parts were locally available. But the PlayPump was proprietary; maintenance was supposed to be handled by the nonprofit organization or contractors, funded by those billboard ads. Except the ads didn't always materialize, and the maintenance hotlines went unanswered.

One Mozambican official, Joaquim George of the rural water authority, summed it up: "Once the pump breaks, and it takes more than three months to repair, people in these communities no longer trust the PlayPump because they are demoralized." In other words, the very people the PlayPump was supposed to help ended up worse off, demoralized, and back to square one (or worse, because sometimes the installation had removed a functioning hand pump to put in the PlayPump).

This scenario is a vivid illustration of a phenomenon that development professionals ruefully call "the playground that nobody plays on." It is what happens when outsiders impose a solution that does not align with local context or preferences. The idea may be novel and even effective under certain conditions, but if it doesn't acknowledge the day-to-day reality of the users, it will be underutilized or abandoned. In the case of PlayPump, a significant oversight was the failure to consider human behavior and culture. The design assumed unlimited play. But in many communities, play is a luxury when there is work to be done, and the boundary between "play" and "child labor" can blur if children are essentially needed to operate critical infrastructure. It also assumed children's play energy could be harnessed at scale without considering fatigue and interest. Furthermore, community ownership was lacking. The PlayPump arrived as a donation-free gift, so villagers weren't involved in financing or decision-making. Psychology and experience in development projects show that when communities contribute to or choose a project, they feel invested in maintaining it. When something is suddenly introduced into their midst, it can become a white elephant.

The Fall of the PlayPump Hype

By 2010, only a few years after the big push, the narrative had flipped. Investigative reports and evaluations surfaced "troubled water," to quote the PBS Frontline follow-up show's title. PlayPumps International, the U.S.-based nonprofit set up to popularize the pumps, quietly scaled back its ambitions. The Case Foundation (Steve and Jean Case's philanthropy) and other backers had to confront the mismatch between their rosy expectations and reality. To their credit, some funders publicly acknowledged the shortcomings. The Case Foundation released a statement highlighting lessons learned and emphasizing the importance of pilot testing and listening to communities before rushing to scale an unproven technology.

In retrospect, the red flags were there early on, had anyone paused to look. For example, one internal report (commissioned by the Mozambique government) that never initially went public had cited exactly these issues: difficult for women to use, maintenance problems, and kids not playing as expected. But at the height of the excitement, those warnings were downplayed or ignored. Confirmation bias played a role—the stakeholders saw what they wanted to see. Donors visited a fresh installation where kids happily spun on a new merry-go-round, water gushed, and everyone smiled for the camera. It's natural to be moved by such a scene. But that snapshot in time wasn't the full story.

The PlayPump saga underscores something fundamental: True innovation is not just about a clever product; it's about a solution fitting into a human system. That means understanding the daily routines, motivations, and constraints of the people who will use it. It means having empathy and humility to involve those people from the start. In designing ethical solutions—whether for customers, citizens, or in this case, communities—a common pitfall is falling in love with the technology or concept and forgetting to deeply inquire into the needs and wisdom of the end users. The engineers and executives behind PlayPump were not malicious or foolish; they genuinely wanted to help, and they created something inventive. Their failing was one of omission: not incorporating local voices and expertise.

Listening as a Key Leadership Skill

Listening sounds simple, but as this case shows, it's often neglected in the rush to deliver results. True listening in a leadership or innovation context means actively seeking out input from all stakeholders, especially those who might be less powerful or visible. In corporate settings, this could be the frontline employees or customers; in public policy, the citizens or service users; in development projects, the community members. It means asking questions before deciding solutions, and treating those answers as vital data, not as obstacles to a great idea.

Suppose the PlayPump team had conducted a thorough community consultation. In that case, they might have discovered, for example, that women preferred hand pumps due to their familiarity and the ability to control the effort. Or maybe they would have learned that kids primarily play in early morning or late afternoon, which wouldn't yield enough water for all-day needs. With those insights, they could have adjusted the approach, perhaps by keeping the hand pump alongside as a backup, training locals in maintenance, or incorporating a way to use adult leverage on the roundabout. Or, they might even have pivoted to a different water solution entirely for certain areas (since PlayPump only worked where groundwater was not too deep and demand was moderate).

Another layer of listening is monitoring a project over time and collecting feedback after implementation. Failures will happen; the key is to catch them early and adapt. In the PlayPump case, once reports of problems trickled in, a more responsive organization could have paused new installations and fixed issues, such as improving the merry-go-round gear ratio to pump more water per rotation or adding a simple crank handle for adults to use when kids aren't around. But such adaptations require acknowledging feedback without defensiveness. For a while, PlayPumps International stayed publicly optimistic, likely because admitting major issues might jeopardize funding or reputations. It's an understandable human reaction, but a costly one. In contrast, an example of ethical consideration and positive course correction can be found in some humanitarian organizations that deploy technology: when early users report a bug or flaw, they stop the rollout, refine the design, and relaunch properly. The humility to do so can save a project.

Voices of others in the development community have since cited PlayPump as a cautionary tale. One NGO worker remarked, "It reinforced the mantra: 'Ask the locals. They often know best what would help them.' Had anyone asked women in those villages, they might have said, "We'd rather have two reliable hand pumps than one fancy PlayPump." Another expert wrote, "Good intentions aren't enough; you need good listening and iteration. No technology can succeed if it's a solution looking for a problem, rather than a response to a problem identified by the community."

Reflection: Beyond Development

In the workplace, leaders sometimes roll out policies or tools they think will boost productivity, only to face employee resistance or poor adoption. Often, the reason is the same: lack of input from those who do the day-to-day work. A software system might technically streamline a process, but if it disrupts workers' routines or is implemented without training, it might end up causing frustration and inefficiency. Savvy leaders know to pilot such changes with a small group, gather feedback, make adjustments, and earn buy-in before full implementation. That's ultimately listening—and ethics—in practice.

The concept of "user-centered design" or human-centered design has gained traction in recent years across industries. It formalizes what should be common sense: involve the end user at every stage of the solution creation process. Observe them, interview them, co-create with them, test prototypes with them, and iterate based on their feedback. This approach not only leads to more effective solutions but also ensures a sense of autonomy and relevance.

One heartening epilogue to the PlayPump debacle is that many organizations took its lesson to heart. In some places, after the merry-go-rounds were removed, NGOs and governments went back to basics, installing reliable hand pumps or solar-powered electric pumps (depending on context), but this time doing so with community committees involved to oversee maintenance. The pendulum swung back to a focus on appropriate technology, meaning solutions that match the skills, culture, and needs of local people. Proper doesn't mean low-tech necessarily; it means the right tech for the circumstances. For instance, in some villages, the best solution might be a rope-and-washer pump that locals can fix with spare parts, while in others, a high-tech solar pump is great if there's training and supply chains for replacement parts. The key is matching the solution to the context, which comes from listening to stakeholders and considering a decision's larger systemic impact.

Key Takeaways from Chapter 5:

- **Engage Stakeholders Early and Often:** Before implementing a solution, talk to the people who will use or be affected by it. Their insights can highlight potential issues that outsiders might never foresee and building trust at the outset will set the project up for success.

- **Beware of the Hype Trap:** A concept that sounds wonderful in theory or impresses outsiders might have hidden flaws in practice. Validate ideas through small pilots and real-world testing in the actual context.

- **User-Centered Design:** Adopt a human-centered approach—design with the user, not just for the user. Communicate clearly about how it works and how the final product will be maintained. Involve them in brainstorming, design, testing, and refinement of any product or policy that's meant for them.

- **Establish Feedback Loops:** After rolling something out, actively seek feedback. If problems emerge, address them quickly. Be willing to adapt or even halt a project in light of credible negative feedback rather than pushing unthinkingly ahead.

- **Cultural and Behavioral Fit:** Consider the cultural practices, daily routines, and behavioral incentives of the people involved. A solution that requires people to act against their normal behavior (like playing endlessly or radically changing habits overnight) is likely to fail.

- **Don't Undermine Existing Good Solutions:** When introducing a change, ensure it at least matches or improves upon what people had before. Replacing an existing tool (like a hand pump) with an unproven new one can leave people worse-off if the new one fails.

In summary, the PlayPump story is a humbling reminder that listening is not a passive act; it is an active, strategic component of ethical innovation and leadership. When we fail to listen, we risk solving the wrong problem or creating new ones. But when we do listen—truly and earnestly—we increase the chances that our efforts will genuinely improve lives and be embraced by those they're meant to help. As we move to the next chapter, we shift from a story of a failed project due to a lack of listening and adaptation, to a tale of companies that faced a need to adapt to changing times.

Chapter 6:
Adapt or Die – Lessons from Kodak vs. Fujifilm

In the mid-twentieth century, the name Kodak was practically synonymous with photography. The phrase "Kodak moment" entered the lexicon to describe a picture-worthy instant. Kodak film and cameras dominated the global market; the company was an icon of American innovation and success. Meanwhile, across the ocean, Fujifilm (Fuji Photo Film Co.) was a formidable competitor but still often seen as the distant number two, especially in the U.S. market. For decades, these two companies thrived as the twin pillars of the analog photography industry. Yet when the winds of technological change blew—the advent of digital photography—their fates diverged dramatically. Kodak, the once unassailable giant, filed for bankruptcy in 2012, reduced to a shadow of its former self. Fujifilm, on the other hand, not only survived but transformed itself into a diversified technology company that continues to prosper. How did that happen? This chapter explores the story of Kodak vs. Fujifilm as a cautionary tale and a success story combined. It illustrates the crucial importance of adaptability, strategic foresight, and willingness to reinvent oneself in the face of disruptive change. This willingness to adapt is ultimately a key part of building resilience. It also shows the consequences of refusing to adapt and how complacency and clinging to an old business model can lead even the mightiest company to fall.

Dominance in the Era of Film

To appreciate the gravity of Kodak's fall and Fuji's rise, let's rewind to their heyday. Founded in 1888 by George Eastman, Kodak essentially invented amateur photography with its simple slogan: "You press the button, we do the rest." Kodak's business model was brilliant: sell inexpensive cameras and make money on the consumables (film, paper, chemicals) that people would repeatedly buy to capture and print their photos. By the 1970s and 1980s, Kodak had roughly 80% of the market share in the U.S. for film and a similar dominance in photo printing supplies. It operated like a well-oiled machine, churning out film by the mile and enjoying fat profit margins from its chemical and paper businesses. This model has often been compared to Gillette's razor-and-blade model (give away the razor, make money on blades) or inkjet printers (cheap printer, expensive ink). Kodak didn't quite give away cameras, but it did sell them at low margins to drive film sales. They even introduced the first mass-market disposable camera to get more people to use film (interestingly, Fuji pioneered the disposable 35mm camera in 1986; Kodak followed in 1988).

Fujifilm, a Japanese company founded in 1934, competed vigorously with Kodak, especially in Asia and Europe. Over time, Fuji developed film and photo products that were as good as Kodak's, often at lower price points. In Japan, Fuji was the market leader; globally, they steadily chipped away at Kodak's share through the 1980s and 90s, sometimes igniting price wars. Both companies, however, enjoyed the comfort of a growing pie through most of the twentieth century—people around the world kept taking more photos and thus buying more film. The profits were generous: in the late 90s, film and related products accounted for the majority of both companies' revenues and an even larger portion of profits.

During this analog era, both Kodak and Fuji built formidable expertise in chemical coatings, precision manufacturing, and Research & Development (R&D). Film is a complex product, with dozens of micro-thin layers of light-sensitive emulsions and chemicals precisely engineered. The barriers to entry were high; only a handful of companies could do it at scale. Life was good. But, as the CEO of Fujifilm, Shigetaka Komori, later reflected, "A peak always conceals a treacherous valley." For Kodak and Fuji, that valley was the digital revolution.

The Digital Disruption

The concept of digital photography—capturing images using electronic sensors rather than chemical film—can actually be traced back further than many people realize. Ironically, Kodak itself invented the first digital camera prototype in 1975. An engineer named Steve Sasson at Kodak built a toaster-sized contraption that recorded black-and-white images onto a cassette tape. The invention was seen as curious but not immediately commercially viable, given the low resolution and the fact that few people had computers to display digital images. Throughout the late twentieth century, Kodak continued to invest in digital imaging R&D. They amassed patents and even had early professional digital cameras on the market by the 1990s (mostly expensive devices for press photographers).

However, the true disruption hit in the early 2000s. Digital camera technology improved rapidly, and prices fell. By 2003, more digital cameras were being sold worldwide than film cameras. The immediacy of digital enticed consumers-no more waiting for film to be developed, the ability to take many shots at no extra cost (no film to buy), and eventually the convenience of sharing images online. This was a classic case of disruptive innovation: a new technology that initially might not have matched the old one in all aspects (early digital images weren't as high-resolution as film, perhaps) but offered new advantages and improved quickly to surpass the old tech for most uses.

Now, what makes the Kodak vs. Fuji story fascinating is that both companies saw the digital wave coming, yet they responded differently. Kodak wasn't ignorant or lazy; it recognized the threat and made some moves. Kodak's leadership in the 90s and early 2000s often spoke about transitioning to digital. They spent billions on developing digital cameras and even became a top seller of consumer digital cameras in the U.S. by the mid-2000s. Kodak also invested in online photo sharing (they bought a service called Ofoto in 2001, well before Instagram or even Facebook became big). So, one might ask, why did they still fail?

Kodak's fundamental problem was strategic inertia and a miscalculation in their business model adaptation. They understood photography was going digital, but they tried to transition in a way that still protected their core business of printing. They assumed people would want to print their digital photos as much as they printed film photos. The company's strategy emphasized initiatives such as installing Kodak kiosks in stores where people could make prints from digital files, or using Ofoto as a means to order prints online. In essence, Kodak tried to have it both ways: to participate in digital while still shepherding consumers toward printing (which would keep their profitable paper/chemical business relevant). Unfortunately for Kodak, consumer behavior didn't follow that script. As broadband internet and social media grew, people increasingly shared photos electronically and saw less need to print them. By the late 2000s, the volume of prints was in freefall, and Kodak's former cash cow film and paper business became an albatross. The company had a high-cost structure due to its manufacturing legacy and shrinking revenue.

Fuji, on the other hand, took a more ruthlessly adaptive approach. Under CEO Komori's leadership, Fujifilm did two key things: it diversified aggressively beyond photography, and it squeezed efficiency out of its legacy businesses to fund the transformation. Fuji anticipated that film's decline, while inevitable, would be gradual in some markets but precipitous in others. They prepared by investing in new areas where their technological knowhow could be applied.

One fascinating example is cosmetics. It sounds bizarre—what does a film company know about makeup? But Fuji realized that the collagen used in film (as a medium for coating chemicals) had parallels to human skin (which is largely collagen). They had expertise in anti-oxidation (since film degrades by oxidation, and they had to formulate chemicals to prevent that). So they leveraged these to launch a skincare line called Astalift in 2007. They essentially said: We know how to prevent photos from fading; maybe we can help prevent skin from aging. It was a bold move, but it found success in Japan's beauty market. Similarly, Fuji used its coating and precision chemistry skills to get into pharmaceuticals (acquiring a drug company and developing new diagnostic and biotech products). They expanded further into digital imaging solutions, encompassing not only cameras but also the materials required for LCD screens and other related products. Fuji developed a film for LCDs (used in TVs, smartphones, etc.) that became a big revenue source. They also continued their document solutions (Fuji-Xerox joint venture) and expanded their business in that area.

Importantly, Fujifilm recognized that clinging to the old business was a dead end. Komori didn't hesitate to make painful cuts: Fuji slashed thousands of jobs in the film division, closed factories, and scaled down that business rapidly to match the shrinking demand. They used the cash from the good years wisely to fund R&D in new fields. So as film sales plummeted (and indeed they did—worldwide demand for photographic film dropped over 90% between 2000 and 2010), Fujifilm was balancing that loss with gains elsewhere.

Kodak did cut some jobs and tried diversification too, but with far less success. They ventured into making cheap digital cameras and home photo printers, and briefly into consumer inkjet printers. But these markets became brutally competitive (against the likes of Canon, HP, Sony, etc.), and Kodak often found itself late or losing money on hardware. Unlike Fuji, Kodak did not diversify as broadly. They stuck to imaging-centric businesses. One could argue that Kodak's culture was so deeply rooted in traditional photography that they struggled to envision a Kodak that wasn't "the photography company." Fuji's culture, perhaps aided by a more flexible mindset, allowed it to say, "We are a technology company with core competencies in chemistry and imaging; we can apply those anywhere."

There was also an element of decisive leadership vs. hesitant leadership at play. Kodak underwent several CEO changes in the 2000s, trying different tactics, but none managed to execute a turnaround in time. In the 90s, one CEO, George Fisher, invested heavily in digital but arguably not enough to change the fundamental business model. His successors either doubled down on trying to profit from printing or, when finally trying to diversify, found it too late with too few resources left. Kodak's balance sheet became burdened by debt and pension costs, making bold investments harder. Fujifilm's Komori, by contrast, was praised for being proactive and forward-looking. He treated the digital threat as an opportunity to reinvent, not just something to accommodate.

Contrasting Outcomes

By 2012, Kodak filed for Chapter 11 bankruptcy protection. It was a stunning fall for a company that once had a market capitalization in the tens of billions and was a Dow Jones Industrial Average component. Kodak emerged from bankruptcy in 2013 much smaller, focused on commercial printing and film for professional use, having sold off many patents and businesses to pay creditors. The consumer side of Kodak (the film, cameras, photo kiosks, etc.) essentially vanished or was licensed out to other companies as a brand name. People still refer to Kodak's story, usually as a somber example of missing the boat.

Fujifilm, meanwhile, continued to make profits. By the mid-2010s, less than 15% of Fujifilm's revenue came from traditional photographic products. The rest came from an array of sectors: medical systems (like X-ray and endoscopy equipment), pharmaceuticals (they even helped manufacture vaccines with their biotech), document solutions, optical devices, and yes, still some digital photography (Fuji carved a niche in high-end digital cameras, ironically by leaning into retro styling and superb sensors). The company even took advantage of a certain nostalgia. While Kodak stopped making film for consumers, Fuji continued producing some types of film. A surprising resurgence of analog photography (albeit niche) made Fujifilm's "Instax" instant film cameras and films a hit with young people in the late 2010s. So Fuji managed to profit from both the old and new in creative ways.

One telling anecdote: When Kodak was spiraling down, an internal phrase used by some was "We're Kodak, we don't need to do X or Y." That complacency and reliance on past laurels proved fatal. Fuji's ethos could be summed up in Komori's statement: "We will not be like Kodak." He explicitly studied Kodak's missteps and published a book (later translated as "Innovating Out of Crisis") to share the lessons he learned from the experience. Komori noted that one must never stop asking hard questions about one's business model, even when all seems well. He enforced a sense of urgency at Fuji even while profits were still coming in from film, knowing that the profits were temporary.

It's worth noting that not everything was smooth for Fuji either. They had challenges in their new businesses and had to learn industries outside their historical expertise. But their financial strength and commitment to R&D carried them through. They could afford a few failures in new ventures because they spread their bets. Kodak, by contrast, put almost all its chips on trying to monetize digital imaging directly and arguably held on too long to the vestiges of film.

What Can We Learn?

The duel between Kodak and Fujifilm teaches us multiple lessons about adaptation.

First, disruption waits for no one. Technological change can overturn dominant companies quickly. The presence of brilliant engineers or even early awareness of the tech does not guarantee survival—the organization must be willing to reinvent its core operations and revenue streams. Kodak invented the digital camera but didn't reinvent its business around it; Fuji, on the other hand, didn't invent it but adapted its business to a digital world.

Second, it highlights the danger of the "innovator's dilemma," a term coined by Clayton Christensen. Kodak faced a classic innovator's dilemma: digital photography initially had lower profit margins than film and threatened Kodak's cash cow. By fully embracing digital, Kodak would hasten the destruction of its profitable film business. So internal voices were saying, "Let's protect film as long as possible; digital can be a second priority." Fujifilm, not being the top dog globally (though strong), perhaps felt less attachment to the old profit model and more urgency to explore new territory. They were willing to cannibalize their film business for new digital offerings (and all other kinds of offerings) because they foresaw that if they didn't, someone else would. Sometimes leaders must be willing to disrupt their own successful products before someone else does it to them. That requires foresight and courage.

Third, it underscores the value of diversification and leveraging core competencies in new ways. Fujifilm asked itself, "What are we really good at, at a fundamental level?" The answers included: chemical engineering, materials science, imaging, and precision manufacturing. Then they looked for growing markets that needed those competencies. This is a brilliant exercise any company can try when facing a decline in its main market. Kodak, arguably, could have done similarly. Kodak had great chemical know—how, too; perhaps they could have pushed into pharmaceuticals or other materials. In fact, ironically, after bankruptcy, a trimmed-down Kodak did venture into pharmaceutical chemicals in a small way in the 2020s (even being involved in a proposed COVID-19 supply deal), but that was far too late to save old Kodak.

Another takeaway is the importance of listening to the market and data rather than relying on assumptions or hope. Kodak hoped people would print digital photos because that's how they'd always preserved memories. But the data quickly showed declines in printing. There's a parallel with the previous chapter's lesson: listen to stakeholders. Here, the stakeholders are customers and the broad trends of consumer behavior. Kodak had market research but was in denial about the magnitude of change or misinterpreted the data due to wishful thinking. Fujifilm seems to have been more realistic, possibly noticing, for example, that young people in Japan were embracing camera phones early on (the world's first camera phone was in Japan around 2000) and not printing much, sending a signal that the culture around photography was shifting.

Voices of insiders from that era highlight these differences. A former Kodak vice president once explained that by the late 90s, the company was aware that every digital camera sold would reduce film sales, but they still didn't act drastically enough. He lamented that the corporate culture couldn't pivot from its reliance on the profit margins of physical film. Meanwhile, a Fujifilm manager recounts how Komori gathered his team in the early 2000s and said in effect: "We have perhaps ten years to rebuild this company completely. Use our strengths to find new growth. Everything is on the table." That set the tone for radical innovation inside Fuji, even as they continued to profit from film sales to fund the changes. Fuji also spent time looking outward, seeing how markets around them were evolving.

One might wonder: could Kodak have done what Fuji did? Possibly, yes. They had the talent and resources at one point. But it would have needed a mindset shift and more drastic actions earlier. In the 1980s, Kodak diversified into pharmaceuticals by acquiring Sterling Drug, but it sold it off in the 1990s to refocus on photography, which turned out to be the wrong move in hindsight. Fuji kept its diversified ventures rather than overly relying on its core.

Adapting in One's Own Life or Work

This tale isn't just for corporate giants. At a personal level, the principle "adapt or die" applies to careers and skills, too. Many professionals have seen their industries upended by new technologies (think of how travel agents had to adapt when online booking came, or how print journalists had to adapt in the age of digital media). As highlighted in Chapter 2, AI is only the latest in a line of new technology that has disrupted and deeply changed certain industries (and society too). In these cases, those who thrived often were the ones who saw change coming, learned new skills, and repositioned themselves. The ones who assumed "my comfortable status quo will last" usually faced shocks.

One can channel a bit of Fujifilm's approach: identify your core skills and see how they can apply elsewhere if needed. Continue to learn and be curious about emerging trends. Avoid the complacency of thinking, "Because I was successful in the past doing X, I will continue to be successful in the future." The half-life of any skill or business model can be surprisingly short these days. At the same time, Fujifilm's success can be tied back to its trust in its own core abilities as a company.

That doesn't mean constantly change for change's sake-Kodak's brand still stood for something valuable (memories, imaging quality). It means being willing to question and reinvent the delivery of that value. Interestingly, Kodak could have perhaps pivoted to be a memories company in the digital age (possibly leading in online photo sharing and cloud storage of images, etc.). That's essentially what companies like Instagram and Google Photos eventually became, filling the void for consumers that Kodak left. The value of "preserving and sharing memories" never went away; just the means to do it changed. Kodak perhaps didn't fully embrace that new means, whereas new players did. Similarly, we can see how values like integrity and resilience can remain at the core of any endeavor, even if its external application is different.

Key Takeaways from Chapter 6

- **Anticipate and Embrace Disruption:** Technology and markets evolve. Don't ignore trends that threaten your core business—engage with them head-on. If a new technology is going to make your current model obsolete, it's better to be the one reinventing your model than someone else.

- **Be Willing to Reinvent:** Kodak's downfall and Fujifilm's success show the power of reinvention. Sometimes you must redefine what business you're in. Ask "What value do we really provide, and how else could we provide it in a changing world?"

- **Diversify and Leverage Strengths:** Identify your organization's (or your own) fundamental strengths and apply them to new opportunities. Fujifilm turned chemical expertise into new product lines. This diversification hedged against the decline of their legacy business.

- **Overcome Innovator's Dilemma:** Don't let short-term profit preservation undermine long-term survival. It may be necessary to cannibalize your own products with new ones. Better to disrupt yourself than to be disrupted by others.

- **Decisive Leadership:** Strong leadership is needed to drive change. Leaders must cultivate urgency even when current numbers look good. Communicate a vision for the future and take bold actions (cuts, investments, shifts) in time, not when it's too late.

- **Continuous Learning and Flexibility:** The individuals and companies who thrive amid change are those who keep learning and stay flexible. Foster a culture (or personal habit) of curiosity about new trends, willingness to experiment, and comfort with change.

The Kodak vs. Fujifilm narrative reminds us that no company or career is too big to fail if it becomes inflexible. Conversely, with adaptability, even an aging enterprise can renew itself. It challenges us to ask: in our own ventures or professional lives, are we vigilant about changes in our environment? Are we prepared to pivot if needed, even if it means letting go of formerly successful habits? Those questions lead nicely into the next chapter, where we explore an episode that highlights the importance of maintaining trust when navigating a crisis. We'll shift from adaptation in technology to adaptation in response to crisis.

Chapter 7:
Trust in Crisis - The Tylenol Case

One autumn morning in 1982, the world awoke to a terrifying series of headlines: Seven people in the Chicago area had died suddenly and mysteriously after taking Extra Strength Tylenol capsules. Tylenol, a popular over-the-counter pain reliever made by Johnson & Johnson, was a trusted household name at the time, holding over one-third of the U.S. market for analgesics. The notion that a common medicine could turn lethal shook the public to its core. It soon became clear that these deaths were not an accident of manufacturing but deliberate tampering: someone had laced Tylenol capsules on store shelves with lethal cyanide poison. The incident could have become a death knell for the Tylenol brand and severely damaged the reputation of its parent company. Instead, Johnson & Johnson's response to this crisis became a legendary example of ethical leadership and the power of putting customer safety first, even at an enormous short-term cost. This chapter delves into how J&J managed the Tylenol crisis, what guided their decisions, and the long-term payoff of maintaining trust. It is a story of doing the right thing under pressure, showing that in times of crisis, the true character of an organization (or individual) is revealed.

The Crisis Unfolds

On September 29, 1982, the first deaths were reported. A twelve-year-old girl died after a cold, and the puzzle pieces were only put together later when others, seemingly unconnected cases at first, also collapsed the same day in different parts of Chicago's suburbs. Investigators discovered cyanide in Tylenol capsules from the victims' medicine cabinets. It appeared that bottles of Tylenol had been tampered with, likely in stores, since the victims had bought them off the shelf. In total, seven people died in that brief span, including three members of one family. The city was thrown into panic; people rushed to hospitals with fears of poisoning, and authorities urged the public not to consume Tylenol products for the time being.

For Johnson & Johnson, the maker of Tylenol (through its subsidiary McNeil Consumer Products), this was an unprecedented nightmare. Tylenol was J&J's most successful product, responsible for a huge share of revenue and profit in its consumer division. In one swift blow, the brand's association shifted from "reliable pain reliever" to "potential poison." It's hard to imagine a more severe reputational crisis in the consumer goods industry. Moreover, while the company was an innocent victim of a malicious crime, the public's trust in their product was shattered. People often don't parse the nuance in a panic: what they see was: "Tylenol can kill you."

As soon as J&J got wind of the situation, its leadership had to make critical decisions. James Burke, the CEO of Johnson &

Johnson at the time, immediately convened a strategy team. Remember, this was 1982—before the age of instant internet communication, before widespread product recalls were common, and before tamper-proof packaging was standard. It was also a time when corporate crisis playbooks were less established. J&J had to write the playbook as they went along.

Guided by a Credo

What set Johnson & Johnson apart was something intangible but powerful: the company's Credo. J&J had a one-page document, written in 1943 by the founder's son, that articulated the company's core values and responsibilities. It stated, in essence, that the company's first responsibility was to the people who use its products (doctors, nurses, patients, mothers and fathers, and all others who rely on J&J products). It went on to list responsibilities to employees, to the communities, and lastly to shareholders (notably in that order). This Credo was not just a plaque on the wall—it had been integrated into the company's management training and culture over decades.

When the Tylenol crisis hit, James Burke famously turned to the Credo for guidance. In interviews later, he said it was crystal clear: the Credo demands putting customers first, ahead of profits or other concerns. "It made it very clear at that point exactly what we were all about," Burke recalled. "It gave me the ammunition I needed to persuade shareholders and others to spend the $100 million on the recall." That's right—$100 million in 1982 dollars (which would be much more today). That was the estimated cost of what Burke and his team decided to do next: a nationwide recall of every Tylenol capsule product, totaling 31 million bottles. It was an extraordinary step. To be clear, only a handful of bottles in Chicago had been tampered with (to the best of anyone's knowledge), but J&J didn't hesitate to recall all Tylenol capsules from every store, everywhere.

This move astounded many observers. Some said it was an overreaction or corporate suicide; certainly pulling your best-selling product off the shelves would devastate short-term sales and could hand the market to competitors (like aspirin or other acetaminophen brands). There was no precedent for a recall of this magnitude in the consumer goods industry, especially not a voluntary one. The U.S. Food and Drug Administration (FDA) had issued warnings, but J&J went beyond what even regulators were asking at that point.

Burke and the leadership team also immediately warned the public: they went on TV and radio, urging people, "Don't consume any Tylenol product you have if it's a capsule, return it to the store." They set up hotlines for worried customers, offered refunds or exchanges, and cooperated fully with law enforcement in the investigation (which, incidentally, ultimately never conclusively caught the perpetrator; the tampering case remains unsolved).

The reaction from the media and public, which was initially shocked and fearful, evolved into incorporate something else: respect for J&J's transparency and swiftness. Instead of stonewalling or downplaying the incident, the company faced it head-on. Johnson & Johnson did not hide behind corporate lawyers or try to shift blame. They acknowledged the seriousness, expressed condolences to the victims, and framed everything in terms of consumer safety. Trust, as Burke later emphasized, was the key word. "Tell me any human relationship that works without trust," he said. "In the long run, the same thing is true about business."

Rebuilding Trust

A recall alone, however, would not automatically restore the Tylenol brand. Once the immediate crisis was addressed, J&J had to figure out how to regain consumer confidence so that it could sell Tylenol again. Tylenol had plunged from 35% of the market pre-crisis to about 7% after the scare (understandably, people were avoiding it). Many experts predicted that Tylenol was finished. But J&J was determined to bring it back the right way.

They took a couple of months to regroup and introduced Tylenol again with unprecedented safety measures. This included new tamper-evident packaging: for the first time, Tylenol (and soon all similar products industry-wide) would have multiple layers of protection—a foil seal on the bottle, a child-proof cap, and boxed packaging with glued flaps—so that consumers could easily tell if a product had been interfered with. This seems normal now, but those seals and safety features came largely as a result of the Tylenol case, which prompted regulatory changes and new industry norms for tamper-proof packaging.

J&J also offered to swap any remaining Tylenol capsules people had with ones in solid caplet form (which is harder to tamper with than powder in a capsule). They were essentially redesigning the product itself for safety. Moreover, they priced the product low on reintroduction and gave out coupons, making it easier for people to give it another chance.

Another brilliant move in re-establishing trust was a media campaign that acknowledged what happened and explained the new safety steps. They didn't try to pretend it never happened; rather, they communicated, "Here's what we've done to ensure this can't happen again." The company's handling was widely praised in the press. The public took note that J&J put customer safety above profit. It's telling that within a year of the incident, Tylenol's market share bounced back close to where it was before the poisoning. By late 1983, Tylenol had recovered around more than 30% share, an amazing comeback many had thought impossible. This was largely credited to the goodwill earned by J&J's ethical response.

The Tylenol case became a staple in business schools and public relations courses on crisis management. It demonstrated how not to lose trust—which is the foundation of your relationship with consumers. Burke's philosophy was simple: "There's no replacing trust. We could make more Tylenol, we could earn back money, but if people lost trust in our product, it's over." So they safeguarded trust at all costs.

Values as a Compass

At the heart of this story is the role of values and ethical decision-making. Johnson & Johnson's Credo was the value compass that pointed them in the right direction. But having values written down is one thing; acting on them under pressure is another. J&J's leadership did so consistently during this episode. Their response illustrates the concept of stakeholder theory in ethics: J&J prioritized customers (one stakeholder group) and even the general public's welfare above immediate shareholder interests. Notably, the stock price of J&J initially dipped after the news (understandably), but it recovered relatively quickly due to its strong handling. In the long term, the Tylenol brand not only survived but also bolstered the company's reputation. Burke was later hailed as one of the greatest CEOs of the twentieth century and received the Presidential Medal of Freedom, in part for his leadership in this crisis.

One might ask: what if J&J had reacted differently? Suppose they minimized the crisis ("Only a few bottles are affected. It's not our fault. We're not recalling nationwide"). Perhaps there would be fewer losses in the short term financially, but consumers would likely remain wary forever, suspecting that more bottles out there are tainted. If someone else had been harmed later because they left Tylenol on shelves, the damage would have been catastrophic, both morally and financially, via lawsuits. Even without further incidents, people would likely remember that J&J wasn't proactive enough. By doing what they did, J&J enhanced their credibility: they demonstrated, "When we say customer safety first, we mean it."

This engenders loyalty. I recall hearing anecdotes that some consumers were so impressed by J&J's actions that they decided to stick with Tylenol once the scare was over because the company had shown it deserved trust. In effect, Johnson & Johnson turned a potential brand-killer event into a brand-strengthening narrative: one of responsibility and honesty.

Broader Implications of Trust

The lesson of Tylenol extends beyond product tampering. It's about how any organization (or individual) handles a serious mistake or crisis. Mistakes will happen—some beyond your control, some due to your errors. How you respond is what defines you afterward. Do you hide, obfuscate, or blame others? Or do you take ownership, communicate openly, and fix what's fixable? It's often tempting to take the defensive or short-sighted route, especially if there are huge costs involved in "doing the right thing." But Tylenol shows that doing right by your customers and the public ultimately benefits your business. It creates a reservoir of goodwill that money can't easily buy.

In more recent times, we've seen contrasting cases. For example, in the automotive industry, the Toyota recall crisis of 2010 (over reports of unintended acceleration) tested that company's reputation. Initially, Toyota faced criticism for its slow response and confusing messaging, which tarnished its renowned quality image, though it later took stronger action. Compare that with Maple Leaf Foods, a Canadian company that had a deadly listeria outbreak in 2008: their CEO went on TV taking full responsibility, apologizing, and explaining their safety overhaul. The public response was largely positive to how Maple Leaf handled it, and trust was rebuilt. Similarly, in cases of airline crashes or data breaches, companies that step up and communicate candidly tend to recover better.

Voices of others in the business community often reference Burke's handling of Tylenol as the gold standard. Crisis management expert Larry Foster (who was a PR leader at J&J) said the key was being "quick, definitive, and comprehensive" in the response, guided by ethics, not just PR spin. Business ethicist Norman Bowie pointed out that in Tylenol's case, ethics and enlightened self-interest converged: doing the ethical thing also turned out to be the best thing for sustaining the business long term.

There's also a personal leadership angle. Employees at J&J reportedly felt proud of their company's response. It reinforced a culture where people believed in the Credo and thus were motivated to live by it in their daily work. A younger employee in 1982 might think, "If our leaders sacrifice profit for principle when it counts, then these values are real, and I should uphold them too in my sphere of work." Thus, ethical leadership sets the tone for ethical behavior throughout an organization.

Trust, as Burke alluded to, is fundamental not just in business but in any relationship. In a team, if colleagues trust each other, they collaborate freely and share information. If trust erodes, people become guarded, and teamwork suffers. In a community, trust in leaders (or lack thereof) can determine whether people follow guidance during a crisis (imagine public health directives during a pandemic—confidence in the health officials is crucial).

One might wonder: did J&J pay any price for being so upstanding? In the immediate term, yes—the recall and lost sales were a significant impact. But J&J was a sizable company that could absorb it, and they had insurance as well for such events. In the long run, Tylenol regained its position and the company thrived. They turned misfortune into a demonstration of integrity.

Tamper-Proofing and Industry Change

The Tylenol incident also spurred positive change beyond J&J. The whole consumer product industry adopted tamper-evident packaging rapidly. So one could argue the entire market became safer as a result of how this crisis was handled. The FDA updated regulatory standards to require such packaging on medicines.

This shows another aspect: a company's ethical response can set a benchmark that others follow, lifting the bar for all. If J&J had not recalled nationwide or changed packaging, perhaps we wouldn't have safety seals so universally today. But because it was publicized, consumers came to expect it, and competitors had to match it.

In our interconnected world, news spreads even faster, and expectations of transparency are even higher. The Tylenol playbook (swift recall, honest communication, over-deliver on safety improvements) is, if anything, more applicable now. When something goes wrong today—say a faulty smartphone battery causes fires—companies often recall globally and update designs promptly, because they know the alternative (dragging feet, denial) will be exposed and harshly punished by public opinion. And consumer safety is now a collectively upheld value, one that demands a response rooted in ethics.

Key Takeaways from Chapter 7:

- **Put Customers (or Stakeholders) First:** In a crisis, make decisions with the primary aim of protecting and informing those who are affected, even if it means significant cost. Human safety and trust trump short-term financial considerations.

- **Lead with Transparency:** Communicate early, often, and honestly. Admitting the problem and explaining what you're doing about it will earn respect. Withholding or spinning information will likely backfire.

- **Core Values Guide Crisis Response:** If you have a clear set of values or principles, use them as your compass. They can provide clarity in chaotic situations. For J&J, their Credo wasn't just talk; it actively guided their actions.

- **Act Decisively:** In emergencies, delay can be deadly (literally and figuratively). A bold, definitive action (like J&J's total recall) may seem risky, but often it contains the damage and demonstrates leadership. Half-measures can prolong the pain.

- **Think Long-Term Reputation:** Remember that how you handle a crisis will be remembered far longer than the crisis itself. The Tylenol case cemented Johnson & Johnson's reputation for integrity. Conversely, other companies are remembered for botched responses more than for the initial issue.

- **Innovate for Safety:** After addressing the immediate crisis, address the root causes and prevent recurrence. J&J's introduction of tamper-proof packaging not only restored trust in Tylenol but changed industry norms, showing commitment to being part of the solution.

The Tylenol crisis story demonstrates that ethics and good business can align. Doing right by people created a strong foundation for J&J's brand that has endured for generations. Next, we shift gears from crisis response to adapting and shifting organizational culture.

Chapter 8: Empowerment at Work – The Semco Story

Imagine a company where employees set their working hours, choose their salaries, vote on major decisions, and even elect their bosses. A workplace with no formal organizational chart, no HR department policing policies, and hardly any written rules. It sounds radical—perhaps even chaotic—yet such a company still exists (albeit in a smaller form) and has thrived. Welcome to Semco, a Brazilian industrial firm led by Ricardo Semler, who turned traditional management on its head. The Semco story is a fascinating experiment in workplace democracy and empowerment, proving that giving people a voice and stake in their work can lead to astonishing success. Under Semler's unconventional leadership, Semco grew exponentially, sustained high profitability, and weathered turbulent economic times in Brazil-all while offering a level of employee freedom virtually unheard of in companies of its size. This chapter explores how Semco's approach worked, the principles behind it, and what lessons it holds about trusting employees, flexibility, and shared responsibility. Ultimately Semco is an example of how to build a company or organization that can respond adeptly to external change—because flexibility towards internal change has already been built into its foundation.

A Maverick Boss

Ricardo Semler took over his father's company, Semco, in the early 1980s when he was in his early twenties. At the time, Semco was a fairly typical top-down organization making industrial machinery (like marine pumps and food processing equipment). Semler, influenced by ideas of democratic workplaces, began to implement changes gradually. Over the next decade or two, he evolved Semco into what some call a "corporate democracy" or a "self-managed organization."

To visualize Semco's environment: Picture a factory floor in São Paulo where supervisors are not barking at workers. Instead, they know the day's production goals and self-organize to meet them. One lathe operator might decide that today he'll help in assembly if that's where bottlenecks are, without needing permission from a supervisor. Flexibility is key. In Semco's offices, you won't find plush executive suites versus cramped cubicles—in fact, executives don't have permanent offices at all. Ricardo Semler himself gave up his CEO office and often chooses to work in open spaces or even outside by the company's swimming pool (yes, they have one for employees). There is an internal marketplace for job roles. When new positions open, they're posted internally, allowing employees to choose to apply or even split roles between people if they prefer a part-time commitment.

One of Semco's most famous practices is that workers set their pay. How on earth can that function? Well, Semco provides everyone with data on the company's finances, so people see the big picture. When someone (individually or as a team) proposes their salary, they do so in light of transparency about what colleagues make and what the company can afford. The norm that emerged is that people tend not to abuse this freedom-peer pressure and self-regulation keep salaries within a reasonable ranges. If someone proposes an absurd number, social dynamics and shared purpose generally nudge them down. Semco found that people sometimes under-requested raises and had to be encouraged to take them!

Another striking feature: leadership elections. At Semco, employees evaluate their managers and even have a say in choosing who leads them. Managers serve with the satisfaction of their teams in mind rather than that of just higher executives. It's similar to a political democracy: if you're a bad boss, you might not last long because your team's feedback carries weight. This flips the power dynamic and incentivizes managers to truly support and listen to their teams (or else they might be voted out or see their team members transfer away).

Semco has minimal formal hierarchy. Instead of rigid layers, they operate with fluid circles of collaboration. Teams form around projects, and seniority is de-emphasized in favor of ability and interest. The company also eliminated many formal controls. For instance, employees have no timecards or fixed schedules; they decide when to come in, as long as they coordinate with colleagues and meet objectives. They even have a system where all meetings are voluntary-if you think you're not adding value, you can just leave a meeting, no questions asked (forcing meetings to be useful to keep attendees).

The Results: Does It Work?

The skeptic might ask, Okay, all that sounds very utopian, but can a business like this compete in the real world? The answer from Semco's track record: Yes, it can, often outperforming competitors. From the mid-1980s to the late 1990s, Semco's revenues grew by nearly 30% a year on average. Profits followed suit. Despite operating in the Brazilian economy, which went through hyperinflation, currency crises, and wild swings, Semco has managed to adapt quickly. Semler credited their democratic approach for being super agile. With employees empowered to make decisions and change course quickly, the company could respond to external turbulence faster than a rigid hierarchy could. They diversified into new businesses opportunistically, often from ideas bubbling up from the workforce. Semco expanded into areas such as environmental consulting, facility management, and even developed new products like high-tech industrial mixers, moving beyond its traditional pump business. Many of those expansions were driven by employees spotting opportunities and being able to act on them.

Semco also enjoyed extremely low turnover (except in cases where someone didn't fit the culture of freedom, in which case they tended to self-select out). People loved working there-understandably, since it gave them autonomy and treated them as adults. Productivity remained high and, by many accounts, improved because when you trust people, most will strive to keep that trust. Workers weren't sneaking off or slacking just because no boss was breathing down their neck; rather, they felt a personal responsibility to uphold the company's performance that they were helping steer. As one Semco factory worker put it, "There's no covering your ass here.

The intent is to get straight to what needs to be done." Without layers of approval and bureaucracy, people focused on outcomes.

One telling anecdote: In the old days, Semco's manufacturing plant, like many others, had issues with absenteeism-workers calling in sick or not showing up fairly often. After the transformation, absenteeism plummeted. Why? Possibly because people had more control over their time (if they needed a day for personal reasons, they could arrange it without playing hooky) and also because they felt more engaged and accountable when they did set their schedules. It's much harder to resent "management" making you work if you essentially are management collectively.

Financial transparency was another pillar. Semco shares all its financial statements with employees and even teaches everyone how to read balance sheets and profit & loss statements. This demystification of the business made everyone more business-savvy. Instead of workers only knowing their narrow task, they understood how their work impacted costs, revenues, and the bottom line. So, for example, if sales were down one quarter, people would see it and voluntarily might rein in expenses or push to help sales, because they knew what was at stake. It created a sense of ownership typically only found in partners or family businesses, yet here it was among regular employees.

Trust and Flexibility

Semler's underlying philosophy was that trust breeds responsibility, while heavy rules and micromanagement breed petulance or dependence. He often said if you treat adults like children (with rigid rules, constant oversight, little freedom), they'll behave like children—either rebelling or simply doing exactly what they're told and no more. But if you treat them as responsible adults, the majority will step up to that expectation.

He also believed that people crave meaning and self-direction in their work. Give them a chance to shape their work life, and they will be more innovative and motivated. Many corporate environments assume workers can't be trusted (hence time clocks, surveillance, etc.), but Semco proved that if you invert that assumption, you might unlock greater performance.

This radical trust extended to even small things: Semco reportedly has no policy on travel expenses. Employees traveling for work make their own decisions on what is appropriate to spend on flights, hotels, meals-because they're entrusted to treat company money like their own, given they see all the financials. And it works; people don't splurge wastefully, as they feel accountable to their peers. In one instance, when an employee abused the system (staying at expensive hotels far beyond what others did), colleagues confronted him peer-to-peer; it wasn't a manager who had to catch him, the team's culture corrected it.

Of course, Semco's style isn't without challenges. It requires a lot of communication and a certain comfort with ambiguity. Not everyone flourishes in a low-structure environment; a few new hires from traditional companies found it disorienting ("just tell me what my job is!" they'd ask, and Semco would respond, "Figure out where you add value!"). It takes time to adjust, and those who prefer clear directives might self-select out. But those who stay relish it. One engineer initially found it strange to help set his boss's salary and evaluate him, but he soon realized that this approach made the boss treat the team with more respect and cooperation.

The company did maintain some minimal frameworks: for example, profit sharing is practiced, and they have a rule that any one person's pay cannot exceed a certain multiple of the lowest paid worker's pay (to maintain fairness, Semler himself adhered to that multiple for his pay). Major capital investments are discussed in committees that include workers from various levels, ensuring decisions consider diverse perspectives, not just top-down ones.

Wider Impact and Voices of Others

Semco's case became famous through Semler's books *Maverick* and *The Seven-Day Weekend*. It inspired many business leaders to rethink how much rule and control is needed. While not many companies have gone as far as Semco, elements of its model have since appeared in progressive workplaces: flexible hours, flat structures, open-book management (financial transparency), and employee involvement programs.

Charles Handy, a British management guru, praised Semler but also noted that few have tried to copy him because it's "too upside-down for most managers." That quote encapsulates the main barrier: managers often fear losing control. Semco required a CEO (Semler) willing to relinquish control and trust the collective. That's a rare trait because many in leadership got there by making decisions and might have an ego invested in being in charge. Semler's approach only works if leadership truly lets go of the command-and-control mindset.

There was also skepticism: Is Semco's success due to unique Brazilian factors or Semler's personality? Some wondered if it's replicable elsewhere. But we have seen similar principles in other contexts. For example, companies like W.L. Gore (maker of Gore-Tex) have a lattice structure with no formal hierarchy and have innovated well. Valve, a video game company in the US, famously has no bosses and employees choose what projects to work on (with great success in game development). So yes, pieces of Semco's philosophy do work broadly.

Employees at Semco often use the Portuguese phrase *sem jeito*, meaning "without a way" or improvisational. It captures their culture of flexibility and creative problem-solving rather than strict procedures. If a new problem arises, they trust that the people closest to it will collaboratively figure out a solution, rather than escalate through a chain of command.

One Semco employee said in an interview: "At other companies, I had so many meetings about doing things and approvals to get, that actual work was slow. Here, we get together and tackle the issues that need to be addressed. It's incredibly efficient once you get used to it." Efficiency through a less formal process is somewhat counterintuitive—most people think freedom could lead to chaos and inefficiency. But at Semco, freedom is paired with shared clarity on goals and transparent information, which reduces the inefficiencies of bureaucracy and enhances agility.

Lessons in Empowerment and Culture

So what can one take away from Semco's experiment? It underscores that people often have more capacity for self-management and innovation than traditional structures allow. By unlocking that capacity, an organization can benefit from ideas and effort that would otherwise be suppressed. It also demonstrates the power of culture. Semco's culture of trust, participation, and accountability is its "operating system." It doesn't rely on manuals or supervisor oversight; it depends on norms and peer expectations.

This shows that culture can be a stronger and more positive control mechanism than bureaucracy. When everyone buys into common values (like "we're all responsible for the company's success" and "treat each other as adults"), you don't need thick policy books. A brief *Survival Manual* comic book is literally what Semco gives new hires, laced with humor and core principles, instead of a formal code of conduct.

Semco also highlights work-life integration. Semler once noted that if someone wants to go surf on Monday morning because they worked Saturday afternoon, why not? So long as the work gets done, why impose a rigid schedule? This "Seven-Day Weekend" idea was about breaking the artificial separation and monotony of fixed workweeks, allowing people to manage their energy and time more naturally. In today's world, with remote work and flex time becoming more common, Semco was ahead of its time, demonstrating that trust in employees to manage their schedules can be effective.

From a leadership lens, Semco suggests that the role of leaders shifts from controllers to enablers. Semler saw his job as creating an environment (or removing obstacles) for people to perform, not telling them what to do day-to-day. He would periodically leave the company entirely for extended sabbaticals, preventing it from depending on him. In his absence, things ran fine, proving the systems and culture were stronger than any one person. That's a mark of sustainable leadership: build an organization that can thrive without micromanagement or single points of failure. And sustainable leadership is ultimately resilient leadership—Semco has perfected a more agile company because of its grid of mutual trust.

Key Takeaways from Chapter 8:

- **Empowerment Drives Engagement:** Giving employees real authority and voice in decisions (even as far as choosing leaders and setting pay) creates a high level of ownership and motivation—people who have a stake act like stakeholders, not cogs.

- **Trust Your Employees:** Semco's success hinged on trusting workers to do the right thing. Most people, when trusted, strive to prove that trust is well-placed. Mistrust and overcontrol can become a self-fulfilling prophecy, leading to mediocrity or resentment.

- **Transparency is Key:** Open financials and open communication equip everyone to act in the company's best interest. When people see the bigger picture, they can align their decisions with overall goals rather than just their narrow tasks.

- **Flexible Work Practices:** Rigid 9-to-5, one-size-fits-all schedules are not necessary for productivity. Flexibility in work hours and arrangements can yield better work-life balance and boost productivity by accommodating individuals' rhythms and needs.

- **Flatten Hierarchy:** Reducing layers of approval and bureaucracy speeds up innovation and responsiveness. It also encourages leaders to act as coordinators or coaches rather than bosses, which can improve workplace relationships and idea flow.

- **Peer Accountability over Formal Rules:** A strong culture can regulate behavior effectively. When peers have a say in evaluations and when values are shared, people hold themselves and each other accountable. It might even be more impactful than top-down enforcement because it's socially driven.

- **Continuous Adaptation:** Semco's fluid structure allowed it to pivot into new businesses and adjust to economic changes quickly. This agility is increasingly valuable in fast-changing markets. Embracing change and being willing to reinvent roles, teams, and strategies from within can be a competitive advantage.

In summary, Semco teaches that when you unleash human potential in the workplace by removing needless constraints, amazing things can happen. It's a bold reminder that many organizational "rules" are perhaps more habit than necessity. By questioning and redesigning how we work together, we might find better ways that benefit both people and performance. The rapid changes of the twenty-first century demand innovations in company and organizational structure alongside individual adaptation—all in service of honoring autonomy, empowering the individual, and weaving trust into the fabric of a workplace. Having explored empowerment within an organization, let's shift the lens to empowerment (and trust) on a broader social scale in the next chapter.

Chapter 9: Community Empowerment – Microfinance

In a poor village in Bangladesh in 1974, a young economics professor named Muhammad Yunus met a woman who made bamboo stools for a living. Her name was Sufia Begum, and her fingers were calloused from weaving the strips of bamboo. Curious about her situation, Yunus asked how much she earned. What he learned was heartbreaking: Sufia did not have enough money to buy bamboo in bulk, so each time she needed raw materials for a stool, she borrowed about 5 taka (just a few U.S. cents) from a trader. The trader, in turn, required her to sell him the finished stool at a price barely above the cost of the bamboo, leaving her with only 2 taka (mere pennies) per stool in profit. It was a form of indentured servitude; she could never save enough to buy her bamboo stock, leaving her at the mercy of the money-lending trader, effectively trapped in poverty despite being skilled and working hard.

Muhammad Yunus had studied lofty economic theories, but confronting Sufia's plight was a personal epiphany. He realized that a tiny amount of money—the equivalent of a few dollars—could break the cycle for Sufia and others like her by freeing them from predatory lending. He conducted a small survey in the village of Jobra and found forty-two people in similar straits; collectively, they needed only 856 taka (around $27 at the time) to pay off their immediate debts and finance their work independently. On the spot, Yunus reached into his pocket and lent them the $27. This simple act turned out to be revolutionary. All the borrowers repaid him over time, and more importantly, they were liberated from their exploitative creditors. They could sell their goods in the open market and keep the profits.

This was the seed of microfinance, the concept of providing very small loans to impoverished people (often women) to help them start or expand tiny businesses. Yunus went on to found the Grameen Bank (Grameen means "rural" in Bengali) in 1983 to formalize this lending model. Over the ensuing decades, microfinance spread across the globe, empowering millions of people to become self-sufficient entrepreneurs. Yunus's fundamental insight was that poor people are often desperately short of credit, not character or capability. If given access to credit on fair terms, they will use it wisely, work hard, and repay it reliably. This flew in the face of conventional banking wisdom, which saw the poor as "unbankable" (too risky, too costly to serve for too little profit). Yunus and the microfinance movement proved that notion wrong in spectacular fashion, yielding powerful lessons about trust, empowerment, and the ripple effects of giving people a hand up.

The Grameen Model

Grameen Bank's approach was highly innovative. Since poor villagers have no collateral to secure loans, traditional banks wouldn't touch them. Yunus eliminated collateral. Instead, Grameen developed a system of group lending: small groups (usually five people, predominantly women) would come together, and while each person received individual loans for their purposes, the group members supported one another and took collective responsibility. If one member struggled to repay, the others would help or the group might not get future loans until everyone was back on track. This created a kind of social collateral, with peer support and pressure ensuring discipline. But it wasn't pressure in a harsh sense; more often, it was solidarity. The group members wanted each other to succeed because they cared for one another (often as neighbors or friends) and because they all benefited from the group's good standing.

Another counterintuitive choice: focusing on women borrowers. Yunus observed that women in these communities often bore the brunt of poverty; they were the primary ones feeding and raising families, yet they had fewer opportunities and less access to credit than men. He also believed that helping women could have a more significant impact on the whole family's welfare (as many development studies later confirmed—women tend to reinvest in their children's health and education). Initially, some villagers were skeptical of lending to women (cultural norms favored men handling money), but over time, the success of those women silenced critics. Women proved to be astoundingly reliable borrowers; Grameen's repayment rates consistently ran 95% or higher.

Imagine a typical case: A woman named Amina takes a $100 loan to buy a second-hand sewing machine to start a tailoring business. With the machine, she can make garments faster than sewing by hand. She starts selling clothes at the market. As her income rises, she repays the loan in small weekly installments, and soon she has the profit to feed her family better and send her children to school. Perhaps she takes out another loan to expand, such as buying more fabric in bulk at cheaper rates or hiring another woman to help as the business grows. Over a few years, Amina goes from barely surviving to running a micro-enterprise, gaining confidence and respect in her community.

Multiply such stories by millions and you get a sense of microfinance's impact. As of the mid-2010s, Grameen Bank alone had lent over $10 billion to more than 8 million borrowers in Bangladesh. Globally, microfinance institutions (MFIs) of various kinds reached well over 100 million clients, catalyzing countless micro-businesses, from basket weaving in India to grocery kiosks in Kenya to poultry farming in Peru.

Voices of the Borrowers

The transformative effect on individuals is best told in their voices. One Grameen borrower recounted, "Before, I had no say in family matters because I brought nothing to the table. Now, with my little business, I contribute income. My husband listens to me, my children look up to me." There are countless anecdotes of women feeling empowered, not just economically but also socially, as they gain more decision-making power in their homes and villages because they have become earners and even employers.

There's the story of a borrower who started with a loan to buy one cow. She sold milk and repaid the loan. Then she took another trip to buy several more cows, turning it into a small dairy collective with other women. Over time, she built a thriving operation, and her village benefited from the steady milk supply (which improved nutrition) and even some job creation. She eventually sent her children to college, something once unimaginable. She famously said that the initial $30 loan was the "seed" that changed everything—a seed she nurtured into a tree of prosperity.

These stories illustrate the concept of self-employment at the base of the pyramid. Rather than waiting for scarce jobs to trickle down to rural areas, microfinance enabled the poor to create their own careers. In developing countries, formal employment is often limited; many people survive through informal sector activities. Microcredit helped formalize and boost those activities into more sustainable livelihoods.

Skeptics and Challenges

While microfinance has won many accolades (including a Nobel Peace Prize for Yunus and Grameen in 2006), it has also attracted criticism and caution. Not every microfinance venture succeeded; some borrowers defaulted or struggled, which lead to indebtedness. By and large, Grameen had systems to handle that compassionately (rescheduling loans, etc.), but when microfinance spread, not all practitioners were as mission-driven. Some institutions turned quite commercial, charging high interest rates that drew fire for essentially making a profit off the poor (though their counterargument was that high operational costs of tiny loans necessitate higher rates than normal banks, but still remain cheaper than local moneylenders).

There were cases in India, for example, of a microcredit bubble in the late 2000s where overlapping loans from multiple MFIs led some villagers into debt traps. That highlighted the need for careful coordination and client education (making sure they're not borrowing more than they can handle, and ensuring loans are for productive use, not consumption).

Yunus himself emphasized that credit is just one tool; to truly escape poverty, borrowers also need education, healthcare, and fair markets for their products. Grameen expanded into these areas, establishing sister organizations focused on healthcare, education scholarships, and even cellular service (Grameen Phone introduced mobile phones to villages, often financing women to start phone-lending businesses). The holistic approach is crucial.

Nonetheless, the core idea has proven robust: when you extend trust and opportunity to people long excluded from the financial system, many will grasp it and pull themselves up. Yunus liked to say that people with low incomes are like a bonsai tree—which is genetically the same as a tall tree, but when grown in a small pot is stunted. There's nothing wrong with the seed; it's the environment. Give people experiencing poverty the soil (opportunity) they need, and they will grow to their full potential.

The Ripple Effects

Microfinance doesn't just increase incomes; it changes mindsets. It fosters an entrepreneurial spirit at the grassroots. In villages where Grameen operates, one can observe cultural changes: women speaking up more, improved school attendance (since mothers can afford school fees and see the value in education), better housing (loans are also given for building safe houses or sanitary latrines), and even political engagement (some Grameen members have been elected to local councils, bringing women's perspectives to governance).

Economically, local markets became more vibrant. If many people in a village get loans to start various businesses, they generate commerce among each other. One is selling eggs, another tailoring clothes, another running a tea stall-each becomes a customer to the other, circulating money locally rather than all wealth going to an outside moneylender or absentee landlord.

It also instilled a sense of dignity. Charity can sometimes demean the recipient or create dependency, but a loan says, "I trust you to use this and pay it back." It's a hand up, not a handout. Many borrowers express pride that they didn't get a gift; they did it themselves with just that initial boost. After repaying, they often continue growing and eventually no longer needed loans. Grameen now even has many "graduate" borrowers whose enterprises grew enough that they transitioned to normal commercial banks (the irony: those banks now chase customers whom they once shunned after microfinance proved their creditworthiness).

The microfinance concept has expanded beyond business credit: micro-savings and micro-insurance are other innovations, helping the poor build safety nets and resilience. But credit remains a linchpin for enabling investment in income-generating assets.

Another broad lesson: start small, think big. That $27 experiment by Yunus was tiny, but he envisioned that if it worked for one village, why not the whole country? Why not the world? He took that idea and ran with it relentlessly. It reminds us that sometimes solving giant problems (like global poverty) doesn't always require giant solutions from the top-sometimes it's empowering many small solutions at the bottom.

Yunus encountered naysayers who thought giving unsecured loans to illiterate poor women was crazy. Banks told him such people wouldn't know how to use money or would default en masse. He proved the opposite. In doing so, he subtly indicted the mainstream financial system for its biases. He often quipped: "The banks lent me money because I had wealth, but the people who needed money the most got nothing. That's backwards." Microfinance flipped that script: lend to those with nothing but their determination.

Adapting and Expanding the Model

The Grameen model was adapted in many countries. In Latin America, for instance, microfinance took a slightly different form with village banks and credit cooperatives. In Africa, mobile banking (using phones) got integrated, allowing people to transact easily, even far from a bank branch (Kenya's M-Pesa, which we'll explore in the next chapter, further revolutionized access to finance). Wherever it went, microfinance had to adjust to local context (e.g., adjusting loan terms to match agricultural seasons, or addressing cultural attitudes toward women handling money). But by and large, the human capacity to strive and repay held across cultures.

Micro-entrepreneurs around the world share similar smiles when they talk about what their first loan did for them. It's often not just the material improvement, but the psychological boost of knowing someone believed in them when others did not. That faith can awaken their self-confidence. Many women report that joining a lending group gave them a supportive community, an escape from isolation at home, and a platform to learn from others. They'd discuss more than business at weekly meetings—health issues, family challenges, etc. So microfinance institutions inadvertently became community development hubs in some ways.

Reflections on Empowerment

What stands out in the microfinance story, akin to themes we've seen in previous chapters, is the power of empowerment and trust. In the previous chapter, we saw how a boss trusts employees and gets great results. Here, an institution trusts the poorest citizens and gets great results. In both cases, trust is extended to those at "lower" positions in a hierarchy (be it corporate or social). And time and again, that trust is honored.

It also highlights innovation in solving social problems. Microfinance was a social innovation as much as a financial one, challenging the norms of who can be served by finance. It blended business principles (loans, not charity) with a social mission (alleviating poverty). This intersection spawned what we call "social business" or "impact investing" nowadays.

Critically, microfinance alone is not a panacea for all poverty; extreme destitution or crises require other interventions. But it addresses one key piece: the lack of capital for the working poor, which was a huge market failure. By correcting that, it unlocks productivity and improves lives.

Yunus once said his vision was to put poverty in a museum, to make it a thing of the past. Microfinance was one tool toward that dream. Whether poverty will be fully eradicated remains to be seen. Still, tens of millions have achieved better lives via microfinance, and the concept has inspired many other inclusive finance initiatives (like crowdfunding platforms where anyone can lend $25 to a borrower across the globe, etc.). Empowering the poor preserves people's dignity and can bring about real social change. Microfinancing is an example of how a small act of trust can make an enormous social and economic impact. On your own journey, don't underestimate the effects of small acts of trust—whether in employees, colleagues, clients, or stakeholders.

Key Takeaways from Chapter 9:

- **Empowerment over Charity:** Enabling people to earn their way (through credit, tools, training) can be more sustainable and dignity-preserving than one-way aid. It builds self-reliance and confidence.

- **Trust the Unproven:** People who lack formal credentials or collateral often have the drive and discipline to succeed if given a chance. Microfinance succeeded by trusting those whom mainstream institutions wrote off.

- **Women as Change Agents:** Investing in women's economic participation yields multiplier effects for families and communities. Women often use resources wisely to uplift their households, so empowering them can have broad social benefits.

- **Social Innovation:** Creative approaches like group lending can solve problems (like risk management in loans) in novel ways. Innovation isn't just tech gadgets; it can be new processes and models that change lives.

- **Holistic Approach:** Combating poverty requires more than money. Microcredit works best when paired with education, healthcare, and community support. Consider the broader ecosystem within social interventions.

- **Measuring Success in Lives Changed:** Microfinance didn't measure its success by profit (Grameen often barely covered costs) but by how many people left poverty. In assessing any venture, look at the impact on human well-being, not just financial metrics, especially for social enterprises.

The story of microfinance is ultimately one of hope and human potential. It demonstrates that with a bit of empowerment, individuals who have been marginalized can become drivers of their destiny and contributors to economic growth and to their communities. As we continue, we'll explore another story of vision and innovation applied to a pressing global problem and how one man's journey and bold transparency are changing how charity works.

Chapter 10:
Use Technology for Good -
The Charity: Water Story

In 2006, a New York City nightclub promoter named Scott Harrison faced a personal reckoning. He had spent his twenties throwing lavish parties, mingling with celebrities, and indulging in what he later admitted was a morally bankrupt lifestyle. Despite outward success, he felt spiritually empty. Seeking purpose, Harrison left the club scene and volunteered as a photojournalist with a humanitarian mission in West Africa. There, in the villages of Liberia, he encountered the stark reality of people drinking dirty water from swamps and puddles. He met children who were sick, even dying, simply because they lacked clean water—something he, like most of us, had always taken for granted. One encounter stood out: a young girl who walked hours each day to fetch foul water, which Harrison knew would likely make her siblings ill. In that moment, a vision crystallized for him: a world where everyone has access to clean, safe water.

Harrison returned to New York with a burning mission. At thirty-one, he decided to devote himself to solving the global water crisis. But he also carried the savvy of a promoter and the insight that many people in his generation were cynical about charities—often unsure where their donations went or turned off by guilt-driven fundraising. So, he founded charity: water with two radical principles: (1) every penny of public donations would go directly to funding water projects (not overhead), and (2) donors would receive proof of exactly what their money accomplished, through photos, GPS coordinates, and reports from each well or system built. In essence, he aimed to reinvent charity with full transparency, accountability, and compelling storytelling.

This chapter explores how one man's clear vision, combined with innovative engagement of supporters, has brought clean water to over 10 million people in less than two decades. It's a case study in how powerful a galvanizing mission can be when communicated authentically, and how merging creative use of technology with compassion can mobilize a new generation of givers.

A New Charity Model

Charity: water started as a small operation—Harrison's friends and contacts at first—but quickly grew into a global movement. Harrison leveraged his promotional skills to throw fundraising events that felt more like parties than pity-fests. He asked nightclubs to donate venues, fashion brands to contribute goods for auction, and DJs to spin music at charity: water galas. His instinct was that giving can be celebratory and joyful, not merely dutiful or somber.

Crucially, he set up an internal split: a group of wealthy private donors (nicknamed "The Well") covered all administrative costs—salaries, rent, marketing—so that when an average person gave $20, literally all $20 went to a water project. This 100% model was a bold promise to earn trust; it meant charity: water must continually fundraise from two pools (one for overhead, one for projects), but it has paid off in donor enthusiasm. People felt confident giving, knowing none of their money was "lost" to admin.

Furthermore, charity: water embraced technology and transparency. On their website and social media, they shared videos from the field, stories of individuals who got clean water, and allowed anyone to see via Google Maps where each well was built and which community it served. For example, a donor who gave to a campaign might later receive the GPS coordinates of "their" well in Ethiopia, complete with photos of villagers celebrating around the new hand pump. This tangible feedback loop is incredibly motivating; donors feel connected to the impact, rather than as if they tossed money into a black hole. They trusted the organization to follow through on its promises.

Another hallmark of charity: water's approach is optimistic storytelling. Instead of showing only sad, desperate images of kids suffering (the typical charity trope), they focused on the hope and potential unlocked by clean water. Their videos often depict the before and after: the struggle of dirty water, then the joy and smiles when clean water gushes from a new well. This framing attracted many who were weary of traditional guilt-based appeals. As Harrison put it, "We're not trading on shame or pity. We're inviting people to be part of a solution and showing the progress."

Charity: water also brought innovation to fundraising by popularizing the idea of donating birthdays. Harrison himself had asked for donations instead of gifts on his thirty-second birthday, raising $15,000 to build wells in Uganda. He then turned that into a signature campaign: encouraging people to pledge their birthdays—ask friends to donate your age in dollars, for instance, instead of buying presents. This resonated widely; soon kids, teenagers, and adults alike were giving up birthdays to fundraise for water. One famous story was of a nine-year-old girl, Rachel Beckwith, who tragically died in a car accident after starting a birthday campaign. Her story went viral, and tens of thousands of people donated in her memory, raising over $1.2 million, which built 143 water projects in Africa serving 37,000 people. It showed the profound human connection that can come from these personal campaigns.

Impact on the Ground

By design, charity: water works with local partner organizations in each country to actually implement the projects (well drilling, spring protections, filter distributions, etc.). This ensures that solutions are context-appropriate and maintained by communities with support from partners. The range of water solutions vary: deep drilled wells with hand pumps, shallow wells, gravity-fed spring systems, rainwater catchments, bio-sand filters for households, even piped systems with solar pumps for larger communities. The technology chosen depends on geography and need.

The outcomes in communities with new water are transformative. Women and girls, who usually bear the burden of water collection, are freed from hours of walking to distant sources. This time saved can be invested in work or education. Water-borne diseases like diarrhea and cholera drop sharply, improving health and reducing child mortality. One village's story: before, children were often too sick or tired (from hauling water) to attend school; after a well was installed near the school, attendance rose and academic performance improved. The village formed a water committee (often half women) to maintain the pump, which also gave women a respected leadership role. These ripple effects echo across charity: water's projects: better health, more school days, economic opportunities (like small gardens or livestock since water is accessible), and restored dignity.

In Ethiopia, a woman named Tigist recounted how she used to walk six hours a day to a river and back. With a newborn on her back, the journey was grueling and dangerous. After charity: water's partner built a nearby well, Tigist said, "It's like I've reclaimed my life. I have time to cook proper meals, to rest, to care for my baby without panic about water." Multiply that by thousands of villages, and you see why clean water underpins so many aspects of development.

Charity: water's transparency helped share stories like Tigist's directly with supporters. Donors not only see numbers of wells or people served, but hear names and see faces. This humanizes the cause. A supporter from London might watch a video of a well inauguration in Nepal and almost feel present as villagers cheer and dance around clean water. That sense of connection is powerful.

Voices of Supporters

Charity: water's approach attracted an unusually broad base of supporters: from school kids running lemonade stands, to tech entrepreneurs giving millions. Many were first-time donors who said they hadn't given to international causes before but were drawn in by charity: water's spirit and trustworthiness.

One recurring theme from supporters is how much they appreciate knowing exactly what their donation accomplished. A donor might say, "In the past I gave to charities but never heard back. With charity: water, I got an email showing me a photo of a well my money helped build. That made me want to give again—I could see the real difference." It's a feedback mechanism that reinforces generosity. Essentially, transparency became charity: water's brand, which built loyalty much like a customer who trusts a favorite company.

Scott Harrison's own story of redemption is also a voice that resonated. He candidly shares how shallow and selfish he had been, and how serving others saved him from that emptiness. His passion is contagious. Instead of a dry CEO, he's a charismatic storyteller bringing donors along for the journey. At events, he would put dirty water on display, or hand out jerry cans (the heavy yellow containers used in Africa) for people to lift, just to give a visceral sense of the problem. Then he'd show the clean water and the solution's cost (often just $30 can give one person water). It motivated people to think, "For relatively little, I can help transform someone's life."

This model of personal engagement and authenticity is something any leader can learn from: people are moved by genuine passion and clarity of purpose. Harrison turned his life around and channeled it into a mission—that authenticity helped galvanize others. Many volunteers and staff joined because they believed not only in the mission but in the transparent, innovative culture of the organization.

Challenges and Adaptations

As charity: water grew (it has now raised over $500 million and funded more than 60,000 water projects in 29 countries), it faced challenges too. Managing the 100% model is one—needing to cultivate separate donor streams and making sure overhead funding keeps pace to support the expanding project volume. They've done this via a loyal group of high-net-worth donors and grants specifically for operations, but it's a constant effort.

Ensuring the quality and sustainability of water projects is another challenge. Building a well is one thing; keeping it operational for decades is another. The sector as a whole grapples with maintenance—too many wells in developing countries break after a few years and sit unusable. Charity: water has tackled this by investing in remote sensor technology on pumps (to monitor functionality and water flow in real time) and emphasizing training of communities and local governments for maintenance. They've also shifted more to larger piped systems where feasible, which may serve more people reliably than many scattered wells.

Measuring impact is always critical. It's not just about counting wells, but outcomes like health improvement and time saved. Charity: water and peers have increasingly done more rigorous impact evaluations to quantify results. Early on, the moving stories sufficed for donors; as the organization matured, they also leveraged data to show results (like X% reduction in waterborne illness after a project, etc.).

The philanthropic landscape has also evolved with more competition and collaboration. Charity: water has maintained a unique niche while also partnering with coalitions (like The Water Effect) to unify messaging on the water crisis.

COVID-19 was an unexpected test: fundraising events were canceled, and field work slowed. Charity: water leaned on its digital community (it was well-placed as it always cultivated online fundraising) and communicated how vital clean water is for hygiene in a pandemic. They managed to ride through by adapting existing campaigns (virtual events, emphasizing water for handwashing, etc.).

Inspiring a Movement

Arguably, charity: water's biggest legacy is how it inspired a new way of giving. Many other nonprofits took cues: adopting the 100% model (or variations of it), using better storytelling and design to engage donors, and being more transparent about impact. Harrison's background in marketing led him to brand charity: water almost like a start-up—with sleek imagery, emotionally resonant videos, and clear calls to action. This was somewhat revolutionary in the nonprofit world in 2006, which was often seen as behind the times in marketing. Now, it's common for charities to have savvy social media, compelling narratives, and clearer info on where money goes. Part of that shift can be credited to charity: water raising the bar and leveraging technology for its cause.

It also recruited a new generation into philanthropy. Young people who might not have been habitual donors became passionate about raising money for water because it felt concrete and positive. This has long-term implications; once someone has a good experience with giving, they often continue a lifetime of generosity. Many donors have stuck with charity: water for years, some joining a monthly subscription program they launched (called The Spring), where folks give monthly and see ongoing stories of how their funds are used.

Scott Harrison encapsulated his vision in a few words: "Water changes everything." Focusing clearly on one solvable piece (the water crisis) made it digestible for supporters. He would explain the wide-ranging impacts (health, education, etc.), but the solution remained straightforward: get clean water to people. That clarity helped rally people—it's easier to act when you know exactly what needs to be done. The ethical example here is Harrison's commitment to his vision and values and his use of technology to further that vision (through storytelling, social media, etc.).

Key Takeaways from Chapter 10:

- **Innovation in Engagement:** Don't be afraid to break the mold in how you engage the public. Charity: water turned fundraising into a personal, enjoyable, and transparent experience rather than a guilt trip. Creative approaches (like donating birthdays) can open new streams of support.

- **Build Trust through Transparency:** By showing donors exactly where funds go and how they help, you earn trust and loyalty. In any project, consider how you can provide feedback and accountability to stakeholders.

- **Harness Storytelling:** Data is important, but stories move hearts. Sharing personal stories of those impacted (and of the founder's own journey) creates an emotional connection that can drive action. Pair those stories with visuals and media that capture attention.

- **Use Technology for Good:** Embrace tools like social media, mapping, remote sensors, etc., to amplify impact and keep donors connected. Charity: water leveraged tech not just for efficiency but as part of the donor experience (e.g., GPS coordinates of wells).

- **Sustainability and Follow-up:** Achieving the immediate output (a well-built) is only step one; plan for maintenance, training, and measuring outcomes over time. Long-term success requires thinking about who will keep the solution working in the years to come.

- **Culture of Optimism and Action:** Charity: water's tone is hopeful, focusing on solutions rather than lingering on problems. This cultivates a culture where donors and staff alike feel energized and motivated, rather than overwhelmed or depressed by the issue.

Charity: water demonstrates the potent combination of a compelling mission and innovative execution. It reimagined how we solve a fundamental human challenge by bringing clean water to millions and revitalizing the concept of charity itself for many. The lesson for any changemaker is to marry vision with trust-building and creativity. When people truly believe in a cause and see that their involvement matters, they will rally to make the vision a reality. Next, we will examine a different type of collaboration and learning across cultures.

Chapter 11:
Cross-cultural Collaboration - The NUMMI Story

In the early 1980s, one of General Motors' worst-performing auto plants was in Fremont, California. The factory had a terrible reputation: workers were disgruntled, often drunk or high on the job; they engaged in acts of sabotage like installing Coke bottles inside car doors (so they'd rattle later) or intentionally using wrong parts. Absenteeism ran rampant on some days, with a fifth of the workers not showing up, leading to frequent production stoppages. The quality of the cars produced was abysmal. Eventually, GM shut the plant in 1982, laying off all its workers. By all accounts, it was a textbook case of a toxic work culture and manufacturing failure.

Around the same time, halfway across the world in Japan, Toyota had become known for making extremely reliable cars at low cost, thanks to its innovative Toyota Production System (TPS), which emphasized continuous improvement (*kaizen*), just-in-time inventory, and respect for workers through empowering them to stop the production line if they spotted a problem (via the famous "andon cord"). Toyota aimed to start manufacturing in the United

States to expand its global presence and perhaps appease rising protectionist sentiments by creating American jobs. Meanwhile, GM realized it needed to learn how to make small cars efficiently to compete with Japanese imports, which were eating into its market share.

Thus an unlikely joint venture was born: NUMMI (New United Motor Manufacturing, Inc.), a fifty-fifty collaboration between Toyota and GM, using the old Fremont plant. Toyota would manage production and bring its methods; GM would learn those methods and receive half the output (small cars to sell as Chevrolet models). They even decided to rehire many of the same UAW (United Auto Workers union) workers who had been at the plant before, including the union leadership who had been antagonistic to GM management. Perhaps to test whether a change in system and culture could turn the same "bad" workers into excellent ones.

Within a year of reopening in 1984 under Toyota's management, the NUMMI plant was utterly transformed. Absenteeism plummeted to a couple of percent, quality metrics soon equaled Toyota's Japanese plants, and the once-problematic workers became proud, disciplined team members producing top-notch cars (the Chevy Nova/Toyota Corolla). This chapter examines how bridging two very different cultures—Toyota's Japanese lean manufacturing and GM's American workforce and traditions—led to remarkable learning and change. It highlights the power of open-mindedness, mutual respect, and a systematic approach in turning around what was thought to be an intractable situation. It's a lesson in how even deeply ingrained work habits and labor-management relations can shift when both sides are willing to learn from and trust each other.

Culture Clash

Toyota sent a team of trainers to California, and they also brought a few hundred U.S. workers and supervisors to Japan for hands-on experience in Toyota plants. For many American workers, this was eye-opening. At the old GM-Fremont, they had been used to adversarial relations—workers vs. bosses, union vs. management. They often felt disrespected and thus didn't respect the job. In Japan, they saw something different: assembly line workers who took immense pride in quality, who stopped the line whenever even a small defect was noticed (rather than letting bad work pass to keep the line moving). They observed managers soliciting ideas from workers on how to improve the process and implementing them. One Fremont veteran recalled initially thinking, "Why are these Japanese workers so enthusiastic and careful? Is it brainwashing?" But after immersing himself in it, he realized, "They treated us like we mattered. They listened. Of course, we wanted to do a good job!"

At NUMMI, Toyota implemented its practices, including creating teams of workers with a team lead, morning group exercises, and meetings to discuss issues. The famous andon cord, which ran along the line, allowed any worker to pull it if they spotted a problem that couldn't be fixed immediately. Pulling the cord triggered a melody and a flashing light at that station; if the issue wasn't resolved in a short time, the whole line would stop so it could be fixed properly. This was unheard of in American plants; stopping the line was an ultimate no-no, as output was king. But Toyota's philosophy was quality first; they believed that a stop-and-fix approach prevents bigger issues later and fosters ownership. Initially, U.S. workers were reluctant: "Am I going to get fired for stopping the line?" But management encouraged it. When line workers pulled the cord and managers rushed over—not to scold, but to help solve the problem—trust began to form. One worker said that the first time he halted production over a misaligned part, he feared the foreman's wrath. Instead, the supervisor thanked him for catching it, and they resolved it together. That was a pivotal moment—he felt respected and part of a team, not just an expendable piece of labor.

Another major change was an emphasis on teamwork and multi-skilling. In the old plant, jobs were narrow and repetitive, and if something outside your specific task went wrong, you'd shrug and say, "Not my problem." At NUMMI, each workgroup was collectively responsible for a section of the line. They were cross-trained, so if one station was struggling or someone was absent, others could step in. This built camaraderie and flexibility. It also meant no one's job was mind-numbingly monotonous—they rotated tasks, which kept people more engaged and less physically strained.

On Toyota's side, they had to adapt too: working with an American union workforce required compromise. For example, in Japan, the norm was for employees to be versatile across roles; in the U.S., union job classification rules were stricter. Toyota had to learn to navigate these constraints and still implement TPS. They eventually won over the UAW leadership by demonstrating that their system would not result in layoffs or overwork. Even if efficiency gains led to fewer labor hours per car, Toyota redeployed workers to perform quality checks or additional training, rather than cutting jobs, thereby fulfilling a promise of employment stability. This was huge for trust-building: the union saw Toyota as an honest partner who valued workers, unlike their past battles with GM management, which often seemed to blame or cut workers at the first sign of trouble.

True communication also required bridging cultural differences. The Japanese approach to feedback is often indirect and focuses on the process, whereas Americans might be more blunt or personal. Through joint training and time, both sides found a middle ground. Many NUMMI employees recalled that the Japanese trainers were patient and humble, not arrogant or condescending. That humility from a "superior" company made Americans more receptive. Toyota managers often would phrase suggestions as, "Maybe we could try doing this differently; what do you think?" inviting input. Over time, U.S. workers realized these suggestions were usually spot-on and started to adopt the Toyota thinking. Meanwhile, Toyota learned to accept that American workers could be more outspoken-that was okay as long as the core principles were upheld.

One example: At first, Toyota's trainers tried to implement group singing of the company song and calisthenics (a common practice in Japan) to build unity. The Americans found it awkward and cheesy. Toyota recognized this and dropped some of those cultural transplants in favor of methods that achieved unity in a more American style (like daily team huddles with discussions, or occasional team outings). It showed cultural sensitivity and a focus on the underlying goal rather than the form.

Dramatic Results

Within months of reopening, NUMMI was producing cars with near Toyota-level defect rates (which were much lower than typical GM plants). BusinessWeek later wrote that the same assembly line workers who used to produce lemons were now building some of GM's highest quality cars. Absenteeism, which was about 20% at GM-Fremont, dropped to about 2% at NUMMI and remained there. Workers who previously clashed with management began reporting that they enjoyed coming to work; they felt their ideas mattered, their contributions were recognized, and everyone was working towards the same target of making a great car rather than adversaries just pushing numbers.

For example, a veteran named Joe, who had been known as a troublemaker before, became a team leader revered for his problem-solving skills. He said it was the first time in his career that he was asked for his opinion on how to improve the job. He had a ton of practical knowledge, and once given voice, he helped eliminate many recurring issues, saving everyone time and frustration.

NUMMI had another interesting effect: it profoundly changed many participants' mindsets beyond the factory. Some union members who had experienced it began sharing with their UAW colleagues that labor and management could indeed trust each other and find mutually beneficial working solutions. Some became evangelists for lean manufacturing, which is rare since lean is often seen as a management initiative. But when done with respect (like at NUMMI), workers embrace it because it makes their jobs easier and more meaningful (less firefighting of problems, more pride in quality).

One worker famously said, "At Fremont, we used to turn out junk, and we knew it. At NUMMI, we built excellent cars and knew it. That feeling of doing it right—it's a big difference." Pride in work is a powerful motivator that often goes untapped when people are treated as cogs.

Sharing the Lessons

GM, for its part, gained invaluable knowledge from NUMMI on the technical aspects of Toyota's system. They got to see first-hand the kanban cards controlling inventory, the standardized work charts, the problem-solving routines, etc. GM did implement some lean techniques in other plants later (and even opened some new plants with lean layouts). However, GM struggled to replicate the full NUMMI magic in its other facilities, often because the deeper cultural elements weren't fully transplanted. Some GM plant managers and union locals resisted the changes or did them superficially (doing some just-in-time here, some quality circles there, but not empowering workers as much or not fully changing attitudes). It showed that adopting lean tools is one thing; adopting the lean culture of mutual respect and continuous improvement is harder, especially across an organization as large and historically siloed as GM. So, GM benefited but arguably under-leveraged what NUMMI taught.

Toyota, on the other hand, succeeded in learning how to manage an American workforce and proved its system wasn't unique to Japanese culture. NUMMI became the blueprint for Toyota's subsequent wholly-owned US plants (starting with Georgetown, Kentucky, in 1988). Those plants hired mostly new workers but heavily emphasized culture from day one, based on NUMMI's experience. They instituted similar team systems and had cooperative relationships with the UAW (Toyota's US plants mostly ended up non-union, but they still often matched union-level pay and conditions to avoid discontent).

One could say NUMMI was one of the most successful cross-cultural corporate partnerships in history in terms of operational results and knowledge transfer. It wasn't without tensions however—initially, American workers thought Toyota's emphasis on every detail (like keeping the work area immaculately clean, or rigorously following standardized work steps) was overkill, and Toyota trainers were frustrated at times by Americans' independent streak (like not following a process exactly because they had a "better way" individually). But those very tensions led to learning on both sides. Americans understood the importance of consistency and process discipline; the Japanese learned to encourage worker suggestions to improve standards rather than allowing people to deviate quietly.

Interestingly, after a few years, Toyota realized the NUMMI workforce was coming up with improvements beyond what Toyota had in Japan. For instance, a team of American workers rebalanced tasks in their section and found a way to do it with one less person but the same output; instead of laying anyone off, they freed that person to form a "moonshine" team (a small group to tinker on process improvements full-time), which then found even more efficiencies. Toyota took some of these innovations back to Japan. It showed that once engaged, any workforce can innovate—good ideas aren't culturally monopolized.

NUMMI also highlighted the importance of management consistency. Some years into the venture, different GM managers cycled in and out of NUMMI's board and tried to push slight changes (or were less engaged in sharing NUMMI lessons internally). When later GM fell into severe financial woes (2000s), some voices said that if GM had truly internalized NUMMI's way widely, it might have avoided some problems. That's speculative, but certainly GM's troubles were more about business model and legacy costs than factory efficiency by then.

From a human perspective, one of the biggest lessons is that workers are not inherently bad or good—it's largely the system and culture they're put in. The same individuals who produced shoddy cars under one system produced excellent cars under another. This echoes W. Edwards Deming's quality philosophy: put a good person in a bad system, and the system wins every time. Toyota provided a good system at NUMMI, and those same "bad apples" from before became exemplary workers. It's a striking testament to how much leadership and process can influence behavior and outcomes. And another example of how empowering workers and how adaptability—in this case a willingness to learn other ways of doing things—can prove to be transformative.

Key Takeaways from Chapter 11

- **Open to Learning from Others:** GM and Toyota both gained by coming together—one seeking to learn superior methods, the other learning to transplant its culture abroad. Being humble about one's weaknesses and willing to learn "foreign" ideas can lead to tremendous improvement.

- **Culture Change is Possible:** The NUMMI turnaround shows that even entrenched negative work cultures can be reshaped by leadership that genuinely invests in people, sets clear expectations, and supports them with the right processes. It requires consistent reinforcement of new values (quality, teamwork, respect) through daily management actions.

- **Empower the Front Line:** Giving workers authority to stop the line or suggest improvements taps into their knowledge and pride. When people see their ideas implemented and their concerns addressed, they become more engaged and responsible.

- **Teamwork and Trust:** Breaking down the us-vs-them between labor and management and fostering a one-team mentality was crucial. Trust was built by Toyota by keeping its word (no layoffs, listening to workers, helping solve problems rather than blaming). Similarly, the union at NUMMI took a risk by cooperating more, which paid off in better conditions and job satisfaction.

- **Adapt Across Cultures:** While core principles remained, Toyota showed flexibility in how practices were implemented in the U.S. context, dropping or modifying some Japanese norms that didn't fit. Likewise, Americans adapted to Toyota's way, recognizing the value behind unfamiliar practices. Successful cross-cultural collaboration requires give and take.

- **Continuous Improvement Mindset:** NUMMI institutionalized *kaizen* (continuous improvement). Workers didn't just do their job; they were expected to constantly think of better ways. Over time, this created a self-sustaining engine of improvement. Any organization can benefit from empowering all members to continuously find and fix problems.

NUMMI's story leaves us with a powerful optimism: that with the right approach, even the most troubled operations can turn around, and adversaries can become collaborators. It underscores how much people's performance reflects their environment: build a supportive, purpose-driven environment, and people will rise to the occasion. As we near the end of our journey through these diverse cases, a common thread emerges: whether in factories, communities, or companies, authentic leadership, trust, and willingness to innovate or adapt can unlock tremendous potential.

In our final case study chapter, we'll look at resilience and renewal in a broader sense—how societies or organizations bounce back from adversity—tying together lessons from all these cases about learning from failure, listening to stakeholders, adapting to change, leading with values, empowering people, and collaborating across divides to act out of integrity and create more resilient organizations.

Chapter 12:
Resilience and Renewal - Rising Strong from Setbacks

Every success story we've explored so far—whether it's an inventor overcoming failures, a community solving water problems, a company reinventing its culture, or an individual transforming a cause—had a backdrop of setbacks and adversity. Resilience, the capacity to bounce back from difficulties, has been a common ingredient in successful recovery. In this concluding chapter, we synthesize insights on resilience and look at a few emblematic cases where resilience led to renewal: how companies revived after near-collapse, how communities rebuilt after disaster, and how individuals persevered against odds to create change. These reflections tie together the book's themes, reinforcing that whether on a personal, organizational, or societal level, the journey of ethical leadership is seldom smooth—but with grit, adaptability, and learning, setbacks can become springboards to new heights.

Corporate Comeback: Apple's Near-Death and Revival

One of the most famous corporate resilience narratives is that of Apple Inc. Today, Apple is a tech behemoth, but in the mid-1990s, it was on the brink of bankruptcy. After a series of failed products and strategic missteps, Apple's market share and reputation plummeted. In 1997, Michael Dell even quipped that if he ran Apple, he'd shut it down and return the money to shareholders. That same year, Apple brought back its co-founder, Steve Jobs, who had been ousted in the 1980s, as interim CEO. What followed is a case study in focused renewal.

Jobs applied many of the principles we've discussed: he simplified Apple's product line (listening to what customers wanted and cutting the rest), he fostered innovation and embraced change (launching the iMac with its bold design was a statement of Apple's identity), and he sought collaboration (even with erstwhile rival Microsoft, convincing Bill Gates to invest $150 million in Apple and develop Office for Mac, which signaled stability for the market). He also reignited Apple's ethos of excellence, motivating employees with a clear vision of making "insanely great" products. The turnaround took a few years; by the early 2000s, Apple was profitable again and then soared with products like the iPod, iPhone, and other innovative devices. Underpinning this comeback was resilience: Apple's team didn't give up when pundits wrote them off. They learned from past mistakes (like the product flops that came from trying to do too much) and doubled down on strengths (user-friendly design, integrated hardware-software experience). The Apple story teaches that even after a fall, returning to core values and embracing bold new thinking can revive an organization. Jobs often said that being fired from Apple initially was painful. Still, it taught him priceless lessons in humility and clarity, which he applied in the second act—an example of personal resilience feeding corporate resilience.

Community Resilience: Rebuilding After Disaster in Rwanda

On a societal scale, few challenges test resilience like the aftermath of conflict or disaster. Consider Rwanda, a nation that in 1994 went through a horrific genocide, losing perhaps 800,000 people in one-hundred days. The social fabric was torn, and the economy was shattered. Fast forward to today, and Rwanda is often cited as a development success story in Africa, enjoying stability, strong economic growth, and progressive social reforms (like high representation of women in parliament, major improvements in health, and poverty reduction). How did this renewal come about?

Rwanda's resilience was shown through a deliberate process of reconciliation and rebuilding. The leadership under President Paul Kagame emphasized unity and *agaciro* (dignity/self-respect). They instituted local community courts (Gacaca) to deal with genocide crimes in a restorative justice manner, aiming to reintegrate offenders who confessed and apologized. This community-driven truth-telling and forgiveness process helped many survivors and perpetrators find a way to live together again. It was an extraordinary feat of collective emotional resilience, choosing healing over revenge to break a cycle of violence.

On the development front, Rwanda engaged citizens in Umuganda, a traditional practice revived where, on the last Saturday of each month, everyone participates in community service like cleaning streets or building homes. This fostered a sense of shared purpose in rebuilding the nation. Coupled with strategic economic planning (investing in ICT, coffee exports, and tourism), Rwanda slowly reconstructed its economy. Today, it's known for its clean streets, efficient public services, and innovation (drones delivering medical supplies, for instance).

The Rwandan people's ability to recover from trauma and work towards a common future exemplifies resilience on a grand scale. It underscores key aspects: facing the past honestly, creating inclusive systems so everyone feels a stake in progress, and instilling hope through tangible improvements. While challenges remain (no society heals overnight, and Rwanda's political freedoms are debated), the progress so far shows that even from the deepest tragedy, a society can find the strength to rebuild and even transform positively.

Personal Resilience: Malala's Fight for Education

At an individual level, consider Malala Yousafzai, the young Pakistani girl who was shot in the head by the Taliban in 2012 for advocating girls' education. Few would have faulted her for retreating in fear after such a trauma. Instead, Malala survived (miraculously, with world-class medical care and fortitude) and returned even more determined. At sixteen, she spoke at the United Nations, saying, "They thought bullets would silence us, but from that silence came thousands of voices." She turned her nightmare into a megaphone for the global cause of girls' education.

Malala's resilience is remarkable not just physically but mentally. She forgave her attackers publicly, showing maturity and compassion well beyond her years. Rather than hatred or victimhood, she chose empowerment. By seventeen, she won the Nobel Peace Prize, the youngest ever laureate. Now in her twenties, she continues her activism through the Malala Fund, supporting education projects worldwide.

Her story aligns with themes we've seen: hardship refined her purpose (like James Dyson's failures leading to invention). Malala said being targeted clarified how crucial education and rights are, strengthening her resolve. Also, community support was key—her father, himself an educator, supported her voice, and after the incident, an international community rallied around her recovery and platform.

Malala shows that resilience isn't just about bouncing back to normal, but bouncing forward to something greater. Her voice post-recovery became far more influential than before. She embodies "what doesn't kill you makes you stronger" in a deeply inspiring way.

Synthesis: Building Resilient Organizations and Lives

Drawing lessons from these and earlier stories, what are the building blocks of resilience and renewal?

- **Purpose and Vision:** A clear sense of purpose serves as a North Star during dark times. Jobs refocusing Apple, Rwanda envisioning unity and development, Malala championing education—all had a guiding vision that pulled them forward when it would be easier to give up.

- **Learning and Adaptation:** Resilience is not just endurance; it's adaptation. The resilient don't just withstand blows, they change in response. Kodak's lack of adaptability led to its collapse, whereas Fuji adapted and thrived. Individuals like Dyson or Harrison adapted strategies after setbacks and came out better. Organizations should instill an attitude that setbacks are information, not defeat—a chance to iterate and improve.

- **Support Systems and Teamwork:** None of these recoveries happened in isolation. Rwanda's people collectively engaged in reconciliation and NUUMI was an example of cross-cultural teamwork.. A resilient community or team stands together. Leaders must cultivate trust and solidarity before crises hit, so that when they do, people come together rather than fracture. That means building relationships, an empathetic culture, and shared values in good times.

- **Innovation and Creativity:** Crises often force creative problem-solving. After a disaster, traditional methods may not be enough, prompting people to innovate. We saw this in PlayPump's funders pivoting after failure and in Yunus crafting a new banking model for people with low incomes. A resilient mindset is open to new approaches. Encourage brainstorming, empower experimentation (like Semco's culture of allowing employees to change how things are done). That way, when challenges arise, the habit of creativity is already built-in.

- **Ethical Foundation:** Acting with integrity and care builds trust, which is the bank account you draw on in a crisis.

Johnson & Johnson's Tylenol credo guided them through the poisoning nightmare and preserved trust. If you treat stakeholders right in normal times, they'll stick with you in hard times. If not, they'll abandon you at the first sign of trouble. Resilience, therefore, is bolstered by an ethical, people-first approach long before adversity knocks.

- **Psychological Resilience:** At an individual level, fostering traits like optimism (believing things can improve), grit (sustained effort), and self-efficacy (belief in your ability to influence outcomes) is crucial. Leaders like Harrison or Yunus had an almost contagious optimism—they envisioned a better future and convinced others to work for it. Encouraging such mindsets in teams (celebrating small wins, focusing on strengths, reminding them of past overcome hurdles) can build a collective psychological resilience.

- **Preparedness and Flexibility:** Some shocks are unpredictable, but many can be mitigated by scenario planning and flexibility. Companies that survived COVID-19, for instance, were often those that could pivot to remote work or e-commerce quickly. They had systems and cultures agile enough to change course. Investing in slack resources (like a financial cushion, cross-trained staff, multiple suppliers) adds resilience to absorb blows. NUMMI had multi-skilled teams, so a missing worker didn't halt production like in rigid plants. Charities like charity: water built a diverse fundraising base (online, events, monthly donors) so that if one channel faltered, they wouldn't sink.

- **Taking Care of People:** Crises are times of stress. Organizations that prioritize their people's well-being during hardships through clear communication, appreciation, and support (such as counseling and flexibility)—foster loyalty and energy to rebound. For instance, when the 2011 earthquake/tsunami struck Japan, some companies like Toyota paused production not just due to supply issues but to ensure employees could attend to families and communities first. In return, employees went above and beyond to help the company recover once they were ready. Compassion in crisis breeds a resilient response.

In closing, resilience and renewal encapsulate the journey from adversity to achievement. The stories we've traversed—from failing prototypes to revitalized companies, from communities gaining water to cultures of empowerment—each carries the message that within challenges lie the seeds of growth. The key is how we respond: with fear or with courage, with rigidity or with flexibility, alone or together, in despair or with hope.

The tapestry of stories we've explored all weave into a fundamental insight: authenticity, action, and resilience feed each other in a virtuous cycle. Being authentic (to values, to voices of others, to a vision) sets a strong foundation; taking action (through innovation, empowerment, collaboration) drives progress; and resilience (learning and adapting from setbacks) ensures longevity and deeper impact. Whether you are leading a team, a community, or your journey, these principles can serve as guideposts.

As we bring this section to a close, think of your own journey. What challenges are you facing? What changes do you aspire to make in your life, your organization, or your community? Know that the principles illuminated by these real stories are not abstract—they are tools you can apply:

- If you have a dream, articulate it and share it boldly with others, inviting them along.

- If you fear failure, reframe it as experimentation—each try gets you closer if you learn from it.

- If you feel stuck in a negative environment, remember how much culture can change; you can be a catalyst by modeling the behavior and values you wish to see.

- If a project or venture falters, step back and listen—what is the situation telling you? What might you do differently? Engage those affected in crafting a solution.

- If you need others' support, be transparent and authentic—trust people with the truth, and they'll trust you with their effort and resources.

- If adversity strikes, take heart from these stories: the darkest times are often followed by the brightest innovations or growth.

May these stories serve as your guideposts on the map that's been charted by past leaders and innovators. May the lessons of real voices and real cases empower you to face your challenges with courage, craft your solutions with creativity, and pursue meaningful change with the conviction that no setback is final. As the people in these stories have shown, the potential for renewal and success is within each of us—waiting to be unlocked by the right combination of mindset, support, and effort. The next "real voice, real lesson" could very well be yours.

Key Takeaways from Chapter 12:

- **Resilience is Active, Not Passive:** It's not just "bouncing back" by enduring; it's an active process of finding solutions, adjusting goals, and taking initiative to create a new reality after a setback.

- **Draw Lessons from Failure:** Treat failures and crises as learning opportunities. Conduct post-mortems, glean insights, and incorporate changes so that each challenge makes you stronger or smarter going forward.

- **Maintain a Resilient Culture:** Create an organizational culture that views challenges as shared problems to solve, not blame games. Encourage open dialogue in tough times and empower all levels to contribute to recovery plans.

- **Emotional Resilience & Mindset:** Cultivate optimism and hope. This isn't naive positivity, but a realistic belief that improvement is possible and that efforts matter. Celebrate resilience stories (like comeback narratives) within your organization to reinforce that mindset.

- **Flexibility & Diversification:** Be ready to pivot strategies. Keep options open, whether that's multiple product lines, adaptable workforce skills, or scenario plans. Flexibility increases your ability to respond effectively rather than being paralyzed.

- **Transformation Through Adversity:** Often, the "new normal" after a crisis isn't just recovering the old but achieving a new level. Aim not just to survive but to evolve. Many innovations and reforms (be it in a business model or public policy) that were hard to do in good times become possible post-crisis when change is imperative. Use that momentum to make bold improvements.

The journey of change is long and winding. It demands authenticity to navigate true north, action to move forward step by step, and resilience to weather storms. Embrace this journey with all its ups and downs. If there is one final lesson from all these stories, it's that meaningful change is possible-often against great odds—when ordinary people commit to extraordinary perseverance and principle. The voices in this section have shown you what's possible; now it's time to chart your own course and calibrate your own inner compass. Go forth and chart the course with an open mind, a full heart, and an unshakable will. The world needs your journey, too.

Part III:
The Inner Compass

Chapter 13:
Ethical Practice at Scale

Technology ethics is not merely a lofty ideal—it must be translated into the daily practices and decisions of engineers, technologists, managers, and organizations. Great principles mean little if they can't guide concrete actions on a tight deadline or in a high-stakes project meeting. This chapter focuses on operationalizing ethics: the practical tools, methods, and cultural measures that allow individuals and companies to consistently apply ethical values across projects large and small. We will introduce techniques like checklists and decision trees that act as prompts to "pause and reflect" during technical work. We will discuss methods of risk analysis and stakeholder mapping to systematically evaluate impacts. We'll also explore how organizations can build an ethical culture—through training, incentives, and leadership—so that good practice scales from the one conscientious engineer to the entire enterprise and supply chain. Through case examples, we'll see how ethical practices (or lack thereof) have played out in real scenarios, and extract lessons on making ethics a habit rather than an afterthought. All of these will serve as tools—practical ways you can calibrate your own compass towards integrity.

From Principles to Action: The Role of Tools

One of the simplest yet most effective aids for ethical decision-making is the humble checklist. In complex engineering environments, checklists have proven their worth in improving safety—for instance, aviation pre-flight checklists or surgical safety checklists have dramatically reduced errors. Similarly, an ethics checklist can ensure that critical questions aren't overlooked in the rush of development. Such a checklist might include items like:

- **Stakeholders:** Have we identified all the stakeholders of this project/decision? Who will be affected (directly or indirectly), and have we considered their needs and values?

- **Benefits vs. Harms:** What are the expected benefits of this technology, and what are the potential harms or risks? (Both short-term and long-term). Are there vulnerable groups who might be disproportionately impacted?

- **Alternatives:** Have we considered reasonable alternative designs or approaches that might be ethically preferable (e.g., more inclusive, safer, or more sustainable)? Why is our chosen approach the best ethical trade-off?

- **Compliance and Standards:** Does our plan adhere to relevant professional codes, industry standards, and laws/regulations? (Ethics is more than just compliance, but compliance is a baseline—a project flouting safety regulations is a red flag).

- **Bias and Fairness:** Have we evaluated the design or dataset for biases or inequities? Did we test the system with diverse scenarios to see if anyone is unfairly disadvantaged?

- **Privacy and Security:** If the project involves data, are we respecting privacy and securing data properly? Have we minimized data collection to only what's necessary (data minimization) and given users appropriate control or notice?

- **Transparency and Communication:** Do users or affected parties know what they need to know about this system? Are we being open about limitations and risks? (E.g., labeling AI-generated content, or warning that a device has certain fail-states).

- **Long-Term & Systemic Effects:** What are the possible broader consequences if this technology scales or is misused? Could it be repurposed for something malicious? Are there environmental impacts? How will it interact with other systems?

- **Accountability and Oversight:** Who is responsible for outcomes of this system? Have we defined clear accountability? Is there a way for issues to be reported and addressed? (For example, a contact for ethics concerns, or a plan for recall/rollback if a serious flaw is discovered).

- **Gut Check / Values Alignment:** Finally, stepping back—does this project align with our core values and the kind of society we want? If something feels "off" to team members, have we discussed it openly?

Using a checklist like this forces a team to pause in the momentum of development and reflect. It acts as a counterweight to common pressures like tight schedules or groupthink ("Everyone else is doing it, let's just push forward.") For example, consider a software team rushing to deploy a new app feature that can track user locations for some convenience. A checklist would make them stop and ask about privacy and consent—perhaps leading them to implement an opt-in or at least inform users, rather than silently collecting data. Or an AI startup developing a facial recognition tool might, thanks to the checklist, realize they haven't tested it on a diverse enough dataset and thus hold off release until they do, avoiding an embarrassing biased system.

Checklists need not be seen as bureaucratic hurdles; they can be framed positively as quality assurance for ethics, akin to unit tests for code quality. Some companies have even created internal "ethics checklists" or worksheets for product teams. For instance, Google's engineering teams reportedly used a privacy design checklist for every product, asking questions like "How might this feature be misused?" and "How will we inform users about data practices?" The key is institutionalizing that moment of reflection. It's far cheaper (in terms of reputation and retrofitting) to catch a potential ethical issue early in design than to fix it after deployment amidst public backlash. This tool can be equally useful on an individual and organization-wide level.

Decision Trees and Scenario Analysis

While checklists help ensure key considerations are not missed, sometimes leaders face complex go/no-go decisions or dilemmas with branching consequences. In such cases, a structured decision tree or flowchart can guide ethical decision-making. For example, an organization might develop a decision tree for deploying AI features:

- Start with: Is this AI being applied in a safety-critical domain (health, transportation, etc.)? If yes, then: Ensure rigorous validation with domain experts, consider requiring human oversight at first, and possibly get external audit. If not (it's low-stakes), proceed to next question but with somewhat less stringent process.

- Next: Could decisions from this AI significantly affect individual rights or opportunities (employment, legal, financial)? If yes, then: implement explainability and appeal processes, involve an ethics review team, etc. If not, continue.

- Then: Does the AI use personal data or infer sensitive attributes? If yes: make sure privacy safeguards and bias checks are in place; get informed consent if feasible.

- Then: If the AI's recommendation were wrong, what's the worst-case outcome? Branch: if worst-case is someone could be physically harmed or unjustly treated, put strong fallback mechanisms (e.g., human confirmation, automatic shutdown if anomalies). If worst-case is a minor inconvenience, a lighter touch is okay.

- Finally: After going through controls, do we still feel uncertain about the ethical impact? If yes, perhaps escalate to a higher management or ethics committee for a decision; if not, proceed with deployment but with a monitoring plan.

This is just an illustrative outline. The idea is to bake ethical reasoning into the gating decisions of a project. Engineers naturally use decision trees for technical debugging ("If the sensor reading is out of range, do X, else Y"). Here we are doing similar logic but for ethics. The goal is to make ethics an integrated part of the "go/no-go" calculus, not an external veto that might be ignored.

For instance, a medical device company could use a decision tree when introducing software updates to a diagnostic AI. The tree might require checking if the update could change clinical decision thresholds—if yes, that triggers a thorough review and possibly informing regulators or users, whereas a cosmetic UI change might bypass heavy review. In doing so, the company ensures that potentially risky changes cannot slip through without ethical scrutiny.

Scenario analysis is another closely related tool,. Here, teams are encouraged to play out "what-if" scenarios especially around failure modes or misuse. This is sometimes called a "premortem" exercise: assume the project has been deployed and something went horribly wrong —describe what that might be. By envisioning a hypothetical future disaster, the team can identify vulnerabilities in advance. For example, a team building a drone delivery system might imagine: "What if a drone malfunctions and crashes into a crowd?" That scenario would lead them to implement geo-fencing around public events, parachute mechanisms, or strict maintenance routines. Or a social media team might imagine: "What if bad actors use our new livestream feature to broadcast violence or propaganda?" That could prompt them to implement content moderation or emergency kill-switches for streams.

In one real case, a large social network rolled out live video streaming without fully considering this, and indeed had incidents of violent or criminal acts being live-streamed, creating ethical and practical crises for the company. A pre-rollout scenario analysis might have predicted that misuse. The lesson is that creativity in imagining misuse is a core ethical engineering skill—it's the security mindset applied to ethics: think like an adversary or think like Murphy's Law, and then address those concerns proactively. Some organizations formalize this by convening a "red team" to test a new product's ethical robustness—e.g., try to trick the AI, try to find bias, try to repurpose the tool for something malicious—and report issues.

Risk Analysis and Proactive Safety Engineering

Engineers are familiar with risk analysis in the context of reliability and safety (like FMEA—Failure Modes and Effects Analysis—in manufacturing, which examines how each part could fail and what the effect would be). Ethical risk analysis extends this thinking to consider harms not just from component failure, but from design or usage decisions. It asks: what's the probability and severity of different adverse outcomes, and how can we mitigate them?

One helpful approach is to categorize risks along two dimensions: likelihood and impact. This forms the classic risk matrix. For example:

- Low-likelihood + low-impact risks can perhaps be accepted (or have simple contingency plans).

- Low-likelihood but high-impact risks (the catastrophic edge cases) need fail-safes or redundancies because even if rare, you cannot tolerate the impact. (Think of a nuclear reactor control system: the chance of a meltdown-causing event must be driven to extremely low levels, and backup systems are mandatory).

- High-likelihood but low-impact risks should be mitigated if possible to avoid death by a thousand cuts (e.g., a User

Interface that frequently causes user confusion—minor individually, but if it happens to every user it erodes trust significantly).

- High-likelihood + high-impact risks are show-stoppers—the design must be changed to avoid these, or the project might need to be rethought entirely.

Applying this to, say, a consumer software product, might look different from applying it to a bridge, but the mindset is transferable. For a social media platform, a high-impact, high-likelihood risk might be "privacy breach of all users' data" —that clearly demands robust encryption, access controls, monitoring, etc. A lower-impact risk might be "spam content might increase." That's annoying but not life-threatening, yet still one might mitigate it with filters in consideration of user experience.

Another technique is stakeholder risk mapping: for each stakeholder group, list what could go wrong for them and how badly. For example, in an online education platform: students (risk: data leak of personal info, impact: medium; or AI grading error, impact: maybe high if it affects grades unfairly), teachers (risk: being surveilled by system, impact: trust issue), school admins, parents, etc. This ensures not only one perspective (often the company's liability perspective) is considered, but multiple perspectives.

Risk analysis in an ethical sense must include non-technical risks like reputational damage, loss of user trust, or societal backlash. For instance, consider the early-2010s launch of Google Glass (the Alternate Reality eyewear). Technically, one might have assessed risk of hardware failure or safety while wearing. But the overlooked risk was social: people were uncomfortable being around someone who could be secretly recording via Glass. The device sparked a public backlash (wearers were dubbed "Glassholes") and was eventually shelved for consumers. If the team had done an ethical risk analysis with stakeholders, they might have foreseen the privacy and social acceptance issues—perhaps adding an indicator light when recording, or focusing on enterprise uses instead of everyday use. So, ethical risk analysis goes beyond engineering failure to adoption failure due to ethical concerns.

Proactive safety engineering is another aspect: building systems that assume something will go wrong at some point and prepare for it. One example is graceful degradation—if an AI driving system loses sensor input, it should slow down or disengage rather than blindly continue. Or if a cloud service AI detecting medical scans has low confidence in a particular case, it flags a human radiologist rather than giving a wrong result. These are design features that emerge from asking "what if?" at every juncture and planning safe behavior. It aligns with the older engineering concepts of fail-safe and fail-soft designs. Ethically, it's taking care that when failures happen (because they inevitably do in any system), the system's design minimizes harm.

One practice borrowed from cybersecurity is the "kill switch" or remote disable for dangerous situations. For instance, many robotics companies program their robots to enter a safe state if network connection is lost or if they detect anomalous commands, to prevent hacking from causing havoc. Self-driving car tests often have a remote monitoring that can stop all vehicles if one behaves erratically. These measures reflect an ethical stance: protect life and property first, even if it means sacrificing the system's performance or availability at that moment.

Stakeholder Mapping and Involvement

We've touched on stakeholders in both checklists and risk mapping. Stakeholder mapping is worth expanding on because it is crucial for ethical practice at scale. The idea is to explicitly identify everyone (people or groups) who is impacted by or can influence a project, and understand their perspectives and interrelationships. In engineering, teams sometimes fall into the trap of "the client and the end-user" as the only stakeholders. But often there are many more: local communities, employees who will operate the system, regulatory bodies, competitors (who might react), supply chain partners, future generations, and the environment (sometimes considered a silent stakeholder).

A thorough stakeholder mapping might categorize stakeholders into primary (directly affected) and secondary (indirectly or having broader interest like NGOs or media). Then, for each, consider what is at stake for them. This process not only highlights ethical considerations (like: "The local community is concerned about noise pollution from our wind turbines. How can we address that?") but also is strategic, because stakeholders who feel ignored can become sources of project risk (e.g., lawsuits, protests, or simply non-adoption).

For example, in infrastructure projects like building a dam or highway, historically some projects failed or faced massive resistance because communities weren't consulted and were harmed (displacements, environmental damage). Now it's much more recognized that you must engage those communities early—not just as a courtesy, but as essential to ethically and successfully implement the project. Techniques include public hearings, participatory design workshops, and incorporating feedback into the engineering design (e.g., adding sound barriers on a highway after community input, or altering a dam plan to ensure downstream water flow for farmers).

In technology, stakeholder involvement might mean bringing in user representatives or advocacy groups when designing features that affect them. For instance, a company developing software for the visually impaired should involve visually impaired users in the design/testing—it sounds obvious, but it's surprising how often products are built "for" a group without that group's direct input. Similarly, when creating an AI that will be used in judicial sentencing, an ethical approach would be to involve not just data scientists and judges, but also public defenders, civil rights groups, and former defendants to ensure the tool is fair and its use acceptable.

Stakeholder mapping also encourages transparency and dialogue. By acknowledging all stakeholders, organizations can communicate more effectively ("We recognize this change affects X and Y, here's how we're addressing their concerns.") It moves a company from a silo mentality to a more ecosystem mentality—seeing itself as part of a community with reciprocal responsibilities.

One emergent idea in ethical design is the concept of a "data ethics canvas" or "ethical impact assessment" which often includes stakeholder mapping. For example, the UK's National Health System (NHS) when evaluating AI tools created a data ethics framework where teams must fill out a canvas of questions, including who the stakeholders are and how each is considered. This formalizes stakeholder thinking in a way similar to how environmental impact assessments became standard for large projects. We may see ethical impact assessments become routine for tech projects, which would be a positive step in scaling ethical practice.

Building an Ethical Culture in Organizations

No tool or checklist can cover every situation. That's why cultivating an ethical culture is crucial —a set of shared norms and attitudes that make doing the right thing the default, not the exception. When ethical thinking becomes part of "how we do things here," it scales naturally across projects and time.

- **What does an ethical culture look like?** First and foremost, it is one where speaking up is encouraged and rewarded, not punished. Recall the Challenger case—a toxic culture silenced warnings. In contrast, companies with robust safety or ethics cultures celebrate those who flag problems. One practice is having formal channels for raising concerns (an ethics hotline, ombudsman, or internal ethics committee) and ensuring there's no retaliation. But beyond formality, it's about attitude: leaders should openly discuss ethical considerations, invite critique, and thank employees who express dissent on ethical grounds. A manager might say in a meeting, "I appreciate you raising that potential privacy issue. Let's dig into it," which signals to everyone that ethics is taken seriously and not seen as derailing.

- **The tone from the top matters:** If executives and senior engineers visibly care about ethics—for example, including ethical goals in project KPIs (key performance indicators) —it sets a powerful example. Some companies add an ethics-related component to performance reviews or project post-mortems ("Did we live up to our values? Did we encounter ethical issues and how were they handled?")

When ethical outcomes are measured, they are more likely to be achieved.

- **Training and education are key tools.** Regular workshops or scenario discussions can keep employees sharp on ethical issues. For instance, a company might hold an annual "ethics scenario day" where cross-functional groups discuss hypothetical dilemmas (like a client asking for something slightly illegal or finding a bias in an AI) and how to handle them. This not only educates but fosters a common language and team ethos around ethics. Role-play exercises can be effective too (e.g., simulate a situation where an engineer discovers a safety flaw close to a product launch—how should they proceed?).

- **Interdisciplinary engagement can be part of culture.** Encouraging engineers to consult with legal, compliance, or ethical advisors is a healthy norm. Maybe every team has an assigned "ethics champion" who liaises with a central ethics office. The goal is to avoid isolation where an engineer might feel a dilemma is theirs alone—instead they know resources exist.

- **Ethics when hiring and giving promotions:** organizations can signal commitment by whom they elevate. If someone who cut corners gets promoted while someone who took a stand gets sidelined, that speaks volumes. But if those who exemplify integrity and thoughtful decision-making are rewarded, it sets an example. Some firms now include scenarios or questions about ethics in interviews—not with a "right answer" in mind per se, but to see if candidates have a process for handling such issues and to emphasize that "we care about this here."

- **Leadership modeling is crucial:** when leaders admit mistakes or seek advice on tough issues, it legitimizes doing so. A story: after the Deepwater Horizon oil spill in 2010, which had at its root a culture problem in parts of the oil industry, some companies doubled down on safety culture. Leaders began meetings with safety moments, sharing even minor near-misses to cultivate an open atmosphere. We can imagine similarly having "ethics moments" in tech teams—e.g., starting a meeting by briefly discussing a news story of a tech ethics failure to learn from others' mistakes.

- **Measurement and accountability:** Some organizations conduct ethics audits or culture surveys. For instance, asking employees annually, "Do you feel comfortable reporting unethical behavior? Do you believe the company would act if something unethical were reported?" These surveys can identify weak spots. External audits can also be considered—just as financial audits check books, an ethics audit might examine if AI systems have bias or if data use complies with privacy promises. It's a nascent idea, but one gaining traction in fields like AI (some companies invite third-party fairness audits of their algorithms).

Case Study: Whistleblower Protections and Cultural Change

Consider the case of an automotive company that discovered some employees were manipulating emissions software (a real scenario analogous to the Volkswagen Dieselgate scandal in 2015). In companies where cheating was allowed to fester, typically the culture had signals that results were prized over integrity. Post-scandal, reforms often include establishing independent ethics oversight, anonymous reporting channels, and explicitly encouraging employees to refuse orders that violate rules. Volkswagen, for example, after Dieselgate, implemented a new ethics and compliance program and fired those involved in the cover-up. They encouraged an internal campaign called "Integrity—it's our duty" to try to shift values. Time will tell how effective it is, but early data showed more employees using internal channels to raise concerns. The cautionary tale for other companies is: don't wait for a scandal. Proactively ensure no one at any level feels that they must break rules to meet targets or hide problems to save face. One early indicator of a robust ethical culture is employees reporting small issues (like minor code of conduct violations or potential conflicts of interest) — that shows they trust the system and take ethics seriously.

Incorporating ethics into design processes is also important. Methodologies like Value-Sensitive Design (VSD) have been proposed in academia, which try to integrate stakeholder values from the get-go. This can involve doing background research on societal values, involving stakeholders in prototyping, and iterating with their feedback. Likewise, Human-Centered Design naturally dovetails with ethics because it focuses on real user needs and pain points, which often brings empathy into engineering.

On a project level, some organizations set up Ethics Review Boards analogous to Institutional Review Boards (IRBs) in research. For example, a tech company might have a panel that reviews high-risk AI projects and gives recommendations or approval before release. This adds a checkpoint and also educates teams on what the organization expects.

Scaling Ethics Across the Supply Chain and Industry

Ethical practice at scale also means looking beyond one's own organization to the broader network. A company can have a good internal culture, but what about its suppliers or partners? Increasingly, firms are held accountable for the practices of those they work with (for example, using a component built with forced labor is an ethical and reputational issue). Therefore, many large companies adopt supplier codes of conduct and auditing. For instance, electronics companies have agreements with suppliers about labor conditions and environmental standards. If a violation is found, they demand improvement or cut ties. This is ethics at scale—leveraging economic influence to raise standards industry-wide.

In software or AI, this may translate to careful vetting of third-party datasets or models to ensure they meet ethical criteria (no invasive data collection, no hidden biases). It could also mean collaborating with industry peers to set voluntary standards, like the Partnership on AI where companies collectively discuss best practices for AI ethics, or the International Electrical & Electronics Engineer's (IEEE) work on ethically aligned design guidelines.

The role of professional societies and education also comes into play: scale is achieved when the average engineer coming out of university already has a grounding in ethics and expects to practice ethically. Thus, pushing for ethics in engineering curriculum and training is part of scaling ethical practice. An engineer who learned about case studies like Challenger or AI bias in school will carry those lessons and perhaps be the one raising a hand in a future meeting to question a decision.

Making Ethics a Daily Habit

Ultimately, the goal is to embed ethical deliberation seamlessly into routine engineering workflows. Much like quality control in manufacturing evolved from a separate step to an ingrained mindset of "quality at every step," we want "ethics at every step."

For example, in agile software development, teams have regular stand-ups and retrospectives. During those, it's feasible to insert a quick ethics check-in: "Any user feedback that raises ethical issues? Any new risks introduced in this sprint?" In retrospectives, discuss if any decisions felt uncomfortable and why. These small additions keep awareness alive.

Some organizations implement "ethical pause" mechanisms—a formal way any team member can call for a pause in deployment if they sense an ethical issue that hasn't been resolved. This is analogous to the Andon cord in manufacturing Imagine a software engineer being able to halt a product launch by escalating a concern to an ethics committee, without career penalty. It sounds radical, but it might prevent disasters. Toyota's system works because any stop leads to immediate problem-solving—similarly, an ethical pause would prompt leadership to address the concern promptly and either resolve it or accept the risk consciously. The existence of the mechanism, even if rarely used, empowers engineers and reinforces that ethical quality is as important as technical quality.

As a concrete example, Microsoft reportedly had an AI ethics review process where sensitive projects (like facial recognition contracts) had to be reviewed by a committee. In 2020, they turned down a few deals (like selling facial recognition to certain police departments) citing ethical reasons, which shows the system had teeth. On the other hand, Google established an AI ethics board but disbanded it after controversy, leading to criticism. The lesson: the structures of ethical boards have to be robust and genuinely supported by leadership, not just PR.

Ethical practice at scale equals the sum of small actions. It's the thousand decisions made by engineers each day—naming a variable honestly, writing a unit test for an edge case, pushing back gently on a manager's unrealistic timeline that might force sloppy work, taking time to document limitations in a user guide—these are not grand policy choices, but they add up to a product that either respects users and is reliable, or one that isn't. If each person in a tech organization acts with integrity and mindfulness of consequences, the overall product and impact will reflect that.

Conversely, if individuals think ethics is someone else's job, things slip through. Therefore, fostering a sense of individual empowerment and responsibility is key. Encourage engineers to see ethics as part of their professional identity ("I am not just a coder; I am a guardian of user trust in my domain."). Professional licensure in fields like civil engineering or medicine creates personal accountability (e.g., a doctor's oath). In software, there is no licensure per se, but companies can cultivate a similar sense by internal norms—for instance, requiring technical staff to periodically sign or acknowledge a code of ethics, as some engineering firms do.

Let's end this section with a brief story: A large e-commerce company once had an issue where a recommender algorithm started suggesting inappropriate products to certain users (due to a quirk in how the algorithm correlated browsing histories). A mid-level engineer noticed this and realized it could offend users and cause media backlash. Instead of ignoring it (since technically it wasn't "broken code"), she informed her manager and they decided to manually tweak the algorithm to filter those cases, even though it slightly reduced immediate click-through rates. When she was later asked why she spoke up, the engineer said she imagined her own family seeing those recommendations and felt it wasn't right. This anecdote exemplifies ethical practice at scale: the company didn't have a specific rule for "don't recommend item X with item Y," it relied on the ethical sensibility of employees and a culture where she felt safe to speak up. And by acting, they likely saved the company from reputational harm and served customers better.

In summary, to practice ethics at scale, organizations must equip their people with tools (checklists, frameworks), processes (reviews, audits, stakeholder engagement), and most importantly, values and culture that encourage everyone to do the right thing consistently. It's about turning ethical reflection into a habit embedded in the engineering lifecycle, much like testing and debugging. When that happens, ethics moves from something that is reactive (cleaning up messes) to proactive (preventing issues and adding positive value). The next chapter will build on this by looking at how leaders can foster and lead such a culture of ethical systems thinking, ensuring that ethical practice not only exists, but endures and thrives through the pressures of business and technological change.

Key Takeaways from Chapter 13:

- **Practical Tools:** Using concrete tools like ethics checklists and decision trees can integrate ethical considerations into daily engineering workflows. They act as reminders to consider stakeholders, potential harms, compliance, and long-term effects before finalizing decisions. Scenario planning and "premortem" analyses help teams anticipate and mitigate possible ethical failures before they happen.

- **Risk and Stakeholder Analysis:** Systematically analyzing risks (likelihood vs. impact) ensures that even rare but catastrophic scenarios are addressed with appropriate safeguards. Involving stakeholders early (through consultations, user testing, etc.) leads to more ethically robust and accepted outcomes.

- **Ethical Culture:** A strong organizational culture of ethics is the best defense against ethical lapses. This means leadership sets the tone by prioritizing integrity over short-term gains, employees at all levels feel empowered to voice concerns, and moral behavior is recognized and rewarded.

When ethics is part of "how we do things," it scales naturally across projects.

- **Embedding in Processes:** Ethics should be embedded in existing processes—from design reviews to testing and deployment. Some organizations have formal ethics or responsible innovation review boards to evaluate high-risk projects, ensuring independent perspectives.

- **Continuous Improvement and Accountability:** Ethical practice is not one-and-done; it requires continuous monitoring and a willingness to improve. Organizations should hold themselves accountable, perhaps via internal audits or third-party assessments, to verify that ethical standards are being met in practice. And since many products involve partners and suppliers, ethical expectations should be extended throughout the value chain (codes of conduct for suppliers, etc.). In short, treat ethical quality with the same rigor as product quality—track it, test it, and never stop refining the processes that uphold it.

Chapter 14:
Toward a Culture of Ethical Systems Leadership

Technical excellence alone will not ensure an equitable and sustainable future. In the face of global challenges and rapid technological change, we need a new kind of leadership—a culture of ethical systems leadership. This means leaders (and indeed contributors at every level) who see beyond their immediate domain to the interconnected webs of people, technology, and environment; leaders who place values like integrity, justice, and public trust at the forefront of innovation. In this chapter, we reflect on why long-term systems thinking and moral clarity are not lofty ideals but urgent necessities. We discuss how leaders can cultivate public trust through transparency and accountability. And we emphasize enduring values—those moral guideposts that remain steady even as we navigate upheaval, whether it's a breakthrough AI or a societal crisis. The goal is to inspire and equip you, the emerging engineer or leader, to champion an ethical vision in whatever systems you touch. The problems of our time—climate change, AI's impact on jobs and privacy, geopolitical strife—are systemic and complex. They demand leaders who are as adept with ethics and empathy as they are with equations and code.

Seeing the Whole System

One hallmark of ethical systems leadership is holistic thinking—seeing problems in their full context, across disciplines and scales. Traditional leadership might focus on optimizing within a silo (e.g., maximizing a company's quarterly performance). Systems leadership broadens the view: how does my company's performance link to supply chain stability, to community welfare, to environmental trends? How does a decision in one domain ripple into others?

For example, consider the transition to renewable energy to combat climate change. A narrow leader might aim to build as many solar farms as fast as possible. A systems leader will ask: Are those farms being built with community consent? What about the supply chain for solar panels are the materials sourced responsibly? How will large-scale solar adoption affect the grid stability at night, and what complementary steps (like storage or demand management) are needed to ensure reliable power for all? In short, the systems leader takes a long-term and interdisciplinary perspective. They recognize that solving climate change isn't just an engineering puzzle; it's also a social, economic, and political one that requires integrating many pieces ethically (ensuring, for instance, that marginalized communities are not left behind in the transition, but rather can benefit from green jobs and cleaner air).

This perspective can be cultivated. It involves deliberately learning beyond one's core expertise: an engineer leader reading up on public policy and sociology, or a tech CEO engaging with environmental science and ethics research. It means inviting diverse voices to planning tables—not just as a formality, but to inform decisions genuinely. Leaders who listen widely often catch systemic issues that others miss.

Systems thinking also implies embracing complexity rather than oversimplifying. A common leadership failure is to ignore inconvenient complexities ("that's not our problem" or "those externalities will sort themselves out"). Ethical leaders do the opposite: they shine light on the hidden coupling and ask, "What are we missing? What unintended effects could this have?" This humility before complexity is critical to avoid narrow solutions that fix one thing but break another. When trust and legitimacy fail in one part (like Congress), it cascades through the whole system. A systems leader in government would anticipate that and build safeguards (e.g., automatic continuing budgets) to prevent such breakdowns. Similarly, in tech, a systems leader might foresee how a social media algorithm geared purely to maximize engagement could erode democratic discourse and proactively adjust course to prioritize healthy communication over raw engagement metrics.

Moral Clarity and Consistency

In turbulent times, people look to leaders for a moral compass—someone who clearly articulates right from wrong and stands by it, even under pressure. Moral clarity doesn't mean being inflexible or dogmatic; it means having well-grounded values and not losing sight of them amid the noise.

For engineers-turned-leaders, one challenge is that the environment can put enormous pressure to compromise on ethics:

competition, financial targets, political constraints, etc. Ethical systems leadership requires the courage to say that some lines will not be crossed. For example, a CEO of a data company might face lucrative opportunities to sell user data in shady ways; to invoke moral clarity would be to firmly say, "We do not monetize user trust at the expense of privacy, no matter the profit." Or an engineering director might be pressured to ship a product that hasn't been adequately safety-tested; moral clarity is pushing back and delaying the launch until it's safe, because the alternative betrays core principles of protecting the public.

My own story, as recounted earlier, shows this kind of clarity: no matter how complex the project or high the stakes, he kept returning to the question, "Does this innovation strengthen or weaken the dignity and resilience of those who depend on it?". If a solution compromised fundamental values or left the vulnerable helpless, I deemed it not true progress. That north-star principle guided me to pause projects that threatened safety and seek counsel in morally uncertain moments. Aspiring ethical leaders can learn from this: articulate your fundamental values (e.g., safety, honesty, fairness, sustainability), and use them as a checkpoint in every major decision.

Moral clarity also involves honesty and integrity in communication. Leaders set the ethical tone by how truthfully they communicate, both internally and externally. During crises, an ethical leader resists the temptation to spin or cover up. Instead, they face facts, admit mistakes, and convey what will be done to fix things. This truth-telling builds trust—consider how the candor of some leaders during the COVID-19 pandemic fostered public cooperation, whereas others' lack of transparency sowed confusion and distrust.

For an engineer leader, consistency is part of clarity. If you preach ethics but cut corners when convenient, people notice, and the culture erodes. Conversely, if you consistently adhere to values—even in small everyday choices, like crediting a team member for their idea (fairness), or refusing to fudge numbers to make a project look on track (honesty)—employees will know you mean what you say. This consistency can sometimes carry personal cost (maybe turning down a bonus tied to unrealistic targets, or losing a client rather than lying to them). Still, it sets a powerful example that inspires others. Over time, such consistency pays back in loyalty and a solid reputation.

Moral clarity often requires moral imagination—the ability to picture the human consequences of decisions vividly. Leaders should constantly remind teams of the real people behind the abstractions. For instance, an AI product team might get absorbed in metrics and forget the end-users; a leader with moral imagination might share a user's story or invite a user to speak to the team, restoring the human perspective. This helps ensure that empathy guides technical decisions.

Earning and Sustaining Public Trust

As we have emphasized, trust is the currency of successful systems. Ethical leaders focus not only on delivering results but on doing so in a way that earns public trust and maintains legitimacy. Trust is built through transparency, engagement, and accountability.

- **Transparency**: People tend to trust what they understand. Leaders should strive to demystify technology and decisions for the public. This could mean publishing plain-language explanations of how an algorithm works and what's being done to prevent bias. It could involve open data sharing, such as when a city shares data on its police AI tool's performance or when a company issues transparency reports on government data requests. During crises, transparency might mean providing frequent factual updates even if the news is bad. One memorable instance is how the CEOs of some companies gave almost daily briefings to stakeholders during the early pandemic about what was known, what was unknown, and what was being done, which helped quell rumors and anxiety. Transparency signals respect—it says, "We consider you, the public, as partners who deserve to know the truth."

- **Engagement**: Earning trust is not one-way; it's a dialogue. Ethical systems leaders actively engage with the communities they serve. This can take the form of public forums, Q&A sessions, participatory budgeting (for civic tech), or user advisory panels. A leader of a social media platform, for example, could hold town hall meetings with users about proposed policy changes. Government or utility leaders might engage the public on major infrastructure plans (similar to how Finland famously involved citizens and experts in developing their nuclear waste storage strategy, which helped build trust in the safety measures). When people feel heard and see their input reflected in outcomes, confidence grows.

- **Accountability**: This is crucial. Mistakes will happen—no leader or system is infallible. The difference is in how one responds. Ethical leaders hold themselves and their organizations accountable. This might mean commissioning independent reviews after an incident and publishing the findings unedited. It could mean compensating those harmed and sincerely apologizing, as well as learning to prevent it from recurring. Leaders who dodge responsibility or scapegoat others lose trust rapidly. On the other hand, leaders who, when faced with a failure, say "This is on us. Here's what we're doing to make it right," often find the public surprisingly forgiving and supportive. People know humans err; what they want is assurance that lessons are learned and responsibility will be taken.

In an age of social media and instant communication, leaders are almost constantly in the public eye. Ethical leadership thus involves a degree of personal integrity and authenticity at all times. A culture of ethical systems leadership isn't just what you do in official capacities, but how you carry yourself generally—because any incongruence (say, preaching sustainability but living lavishly and wastefully, or championing fairness at work but discriminating in personal dealings) can become known and undermine trust.

One should also not underestimate the power of vision anchored in values to rally trust. Great ethical leaders don't just solve immediate problems; they articulate a positive vision that connects with people's deeper aspirations. For example, stating a goal like, "We aim to not only deploy new tech, but to do it in a way that makes our community more inclusive and resilient for generations to come," and backing that with action, can inspire stakeholders to trust and join in. People yearn for leadership that elevates, that appeals to our better angels. By tying technical endeavors to ethical outcomes (improved quality of life, reduced inequity, preservation of nature), leaders can transform skepticism into support.

Enduring Values in Times of Change

Technological transitions and social stressors can test our values. It's easy to hold values in stable times; the real proof comes under strain. As innovations disrupt norms, or during crises like pandemics or conflicts, leaders might feel pressure to abandon certain principles "temporarily." Ethical systems leadership urges the opposite: hold fast to enduring values precisely when they are challenged.

What are these enduring values? While they may be phrased differently by different cultures, some near-universal ones include: respect for human dignity, fairness/justice, responsibility/stewardship, honesty, and compassion. Let's consider how these can guide one's actions during transitions:

- **Human Dignity:** Even as automation and AI reshape labor markets, an ethical leader ensures that those affected (e.g., workers being displaced by robots) are treated with dignity—meaning offering retraining programs, not discarding people as obsolete. In product design, respect for dignity means not designing systems that dehumanize or unduly surveil people. For instance, if implementing workplace monitoring tech, do so in a way that preserves employee privacy and autonomy as much as possible.

- **Fairness and Justice:** In times of change, inequalities often widen (those who adapt easily vs those who don't). A leader centered on justice will try to level the playing field. For example, as digital services expand, ensuring rural or low-income communities get access (bridging the digital divide) is about fairness. Similarly, if an algorithm tends to privilege one group over another, leaders must intervene to restore equity. Justice also implies intergenerational fairness—not solving today's problems by deferring costs or harm to future generations (as climate change has taught us).

- **Responsibility and Stewardship:** This value reminds leaders to be caretakers of the assets and people under their charge, and of the planet. In periods of frantic growth or crisis, it's tempting to exploit resources recklessly. Stewardship tempers that question: are we leaving things better (or at least not worse) for those who come after us? A company dumping toxic waste to cut costs may profit briefly, but fails miserably in stewardship. Ethical leadership takes a long-term perspective: a sense of being a guardian of something precious—be it a community's trust, an institution's integrity, or Earth's ecosystems. As my own career story shows, technological leadership is a means to protect and empower people (e.g., building resilient grids so no one is left literally in the dark).

- **Honesty (Truth-seeking and Truth-telling):** In the era of big data and propaganda, leaders rooted in truth stand out as beacons. Embrace evidence, even if it contradicts your prior assumptions or desires. During the development of a complex system, if testing reveals a flaw, honesty compels you to acknowledge and address it, rather than bury it. This science-based integrity is crucial in fields like public health (leaders trusting the data about a vaccine's safety and communicating it transparently) or climate policy (not downplaying scientific consensus for political expedience). Without truth, decisions become unmoored and public trust vanishes.

- **Compassion:** Often undervalued in technical fields, compassion—the ability to empathize and care—is a guiding light in ethical leadership. It encourages inclusive design (considering those with disabilities and those in poverty) and humane policies (such as supporting employees' well-being, not just extracting work from them). In crises, compassion from leaders—showing concern for people before profit—can define their legacy. For instance, some companies continued paying hourly workers during pandemic shutdowns, even when not legally required, out of a compassionate ethos, which employees and customers remembered.

How do leaders keep these values front and center? It can help to have a personal mission statement or the organization's core values displayed prominently and referred to in decision-making. Some leaders literally carry a card in their wallet with the company values or their personal ethical commitments, to remind them when tough choices loom. Rituals can reinforce values too: starting meetings by highlighting a company value and recognizing someone who exemplified it recently, for example.

Another strategy is learning from history and role models. Reflect on figures who upheld values under pressure—like the engineers who, against odds, insisted on design changes that saved lives. Their stories can be moral touchstones. This book has shared a few; you can seek out more in your field.

Key Takeaways from Chapter 14:

- **Systems Perspective:** Ethical leadership in engineering means taking a holistic view of problems. It requires understanding and considering the interdependence of technological, social, and environmental systems. By looking beyond immediate technical goals to broader impacts, leaders can foresee and mitigate unintended consequences and craft solutions that serve the long-term public interest.

- **Moral Courage and Clarity:** Ethical leaders maintain a strong moral compass, grounded in enduring values like safety, honesty, fairness, and respect for human dignity. They clearly articulate these values and stick to them, especially under pressure. Crucially, such leaders are willing to say "no" to opportunities that violate core principles, and "yes" to the hard right over the easy wrong.

- **Trust through Transparency and Accountability:** Public trust is earned by leaders who are transparent about their actions and accountable for outcomes. When people see a leader consistently telling the truth and acting in the community's best interest, they are more likely to grant trust even in uncertain situations.

- **Enduring Values in Change:** Even as technology and circumstances change rapidly, certain values remain non-negotiable. Leaders ensure that innovations align with human rights, equity, and sustainability. They use new tools to advance these values rather than undermine them. Through crises and disruptive changes, they hold fast to the principle that people—their wellbeing and dignity—are the ultimate priority.

- **Collective Ethical Leadership:** Fostering an ethical culture is a shared endeavor. The mantle of ethical systems leadership can be taken up at all levels—not just CEOs, but also engineers, educators, students, and community members. By championing ethical considerations in our own roles and collaborating with others who value integrity, we contribute to a wider culture of responsibility. The cumulative effect of many individuals leading by example—asking the ethical questions, making principled choices, and inspiring peers—is a powerful force for positive change in organizations and society. Each person's consistent ethical actions help "bend the arc" of technology toward justice and the public good.

Chapter 15 - Letters to Practitioners

In this chapter, we move from a discussion of values and practical tools to the lived experience of young professionals. The letters in this section are based on real situations faced by people at the start of their careers. They remind us that ethics is not an academic exercise—it is a conversation with a colleague, a pause before a release, a choice to speak up when silence is safer. As you read, imagine how you would respond and what resources you would draw upon. Which path is the way of integrity in each situation?

Letter to a Young Medical Innovator

Dear Innovator,

You stand simultaneously at the cutting edge of medical technology and at a personal crossroads. You wrote to me about the startup you co-founded, which is developing a wearable device to monitor heart rhythms. Your excitement was palpable; so was your worry. In your last trial, you discovered a subtle flaw in the device's algorithm. Under certain conditions, it can fail to alert the user to an arrhythmia. The issue didn't show up in the small pilot study until one patient's data revealed the gap. Now your investors are pushing to fast-track FDA approval, glossing over this "edge case." You're torn: fix the problem (which could mean months of delay and extra cost) or push ahead and hope for the best. You asked me, as your former professor, what you should do.

I won't give you a simple answer—you already know the answer. Instead, let me share a story and a principle. First, the story. A few years ago, a prominent blood-testing startup, Theranos, collapsed spectacularly. Its charismatic founder promised revolutionary diagnostics, but behind the scenes, the technology didn't truly work. They fudged data and hid device failures. For a while, they fooled investors and regulators, but eventually the truth came out—as it always does. The company's value evaporated, its leaders faced legal consequences, and worst of all, patients were misled about their health. It's an extreme case, but it highlights a non-negotiable in medical innovation: people's lives and trust are at stake. In healthcare, if you know of a risk and you hide it or ignore it, you are gambling with someone's life and with your integrity. You might get lucky for a while, but the cost of being wrong is measured in heartbeats and human lives. No profit or acclaim is worth that.

Now, the principle: "Hold paramount the safety, health, and welfare of the public." This is the very first canon of engineering and biomedical ethics for good reason. It means that in any clash between expedience and safety, safety wins every time. You mentioned pressure from investors—I understand Perhaps they say a 95% success rate is good enough, or that competitors will beat you if you don't rush.

But I want you to play this scenario out in your mind: Suppose you release the device as-is. Six months from now, a thousand or ten-thousand people are wearing it. One of them relies on it to warn of a heart rhythm problem, but it doesn't send the alert in time. Imagine that person suffers a preventable cardiac arrest. Their family will ask, "Did you know this could happen?" How would you answer? The weight of that question is something you must carry when you make this decision. As an engineer and leader, you are accountable—legally and morally—for what you create. In U.S. law, there's even a doctrine (from the U.S. v. Park case) that holds CEOs personally responsible for safety issues in products, even if they delegated tasks to others. They need not prove intent—if you should have known and acted, you're liable. But beyond the law, consider your conscience and the trust of your users.

I remember your passion for this project. You told me you started it because your father collapsed one day and only got help because someone nearby knew CPR. You wanted a device that would ensure no one is alone in a heart crisis. That is a noble goal. Don't betray that mission by cutting corners now. Yes, being ethical might cost you some funding or a few headlines. But it will earn you something far more important: credibility and self-respect. In the long run, that is what sustains a career. Companies that did "the right thing," like Johnson & Johnson, ended up with stronger public trust and enduring success. Those that cheated or rushed at the expense of safety—well, just ask Boeing how it feels about the 737 MAX saga, or ask the former leaders of Theranos. Shortcuts can lead to very long delays in the end, or to endings that no one wants.

Practically, what can you do? I advise you to speak up to your team and investors clearly and firmly. Present the data on the flaw, and your plan to address it. Frame it not as a setback, but as an essential design improvement. Use the communication approach we practiced: define the issue, state the facts, outline options, and make a principled recommendation. For example: "The issue is the algorithm's alert gap. The facts: in 1 out of X cases, it failed to detect arrhythmia in our test. Our options: (1) delay release to fix it, which costs $$ and time but ensures safety; (2) proceed without fixing, which risks a user's life and a product recall; or (3) implement a partial mitigation (explain) with warnings, which reduces risk but not as much as a full fix. Based on the principle of patient safety first and our company's long-term credibility, I recommend we take the time to fix it fully. If we do so, we will need to manage investor expectations and perhaps secure bridge funding, and if we don't, we must prepare for the consequences—potential harm to users and loss of trust." By presenting it this way, you make the ethical choice crystal clear. Investors are businesspeople—remind them that one wrongful death or scandal could sink the company they've invested in. Ethics isn't just morally right; it's a smart business strategy too.

Lastly, I want you to know: being young is not a handicap here. I was under thirty once, working on projects that had life-and-death implications. I remember the knot in my stomach when I had to tell a superior, "No, we can't proceed like this." It's intimidating. But doing it early in your career sets the tone for yourself. Each time you choose integrity, you make it easier the next time. It's like a muscle you strengthen. And people notice—the right people, anyway. You might lose a deal, but you'll gain mentors, colleagues, and employees who want to work with someone they can trust. That is priceless.

In the quiet moments, when no one else is around, you have to live with yourself. Make the choice that lets you sleep at night and face yourself in the mirror. Ten years from now, you will either tell the story of how you stood by your values and built a truly reliable device, or you'll have the memory of yielding to pressure and regretting it. I believe in you; I've seen your integrity in action in the classroom. Now is the time to double down on it. Hold the line. Fix the flaw. Deliver a device that you know will not fail the people trusting it. That is how you will save lives—and perhaps even save your company in the process.

Sincerely,
Massoud

Letter to a Young Cybersecurity Professional

Dear Security Leader,

I recall the day you graduated from our Security Technologies program—you were eager to take on the world of cyber threats and defenses. Now you've written to me with a dilemma that is both uniquely modern and as old as ethics itself. You've joined a fast-growing tech firm, and you've discovered a vulnerability in their system that could expose sensitive customer data. It's not hypothetical; you reproduced the issue in testing. When you brought it up to your manager, you were told to "leave it alone for now." The product is on the verge of a major rollout, and higher-ups fear that fixing the issue now (and disclosing it) would delay the launch and attract negative attention. They suggest that since "no one has abused it yet," it's okay to patch it quietly later, maybe in the next update. You're uneasy—you recognize this reasoning as a moral gray zone, if not outright wrong. You asked me: Should I push back, even if it risks my job? And if so, how?

Cybersecurity often presents these less visible ethical choices. Unlike a bridge with cracks or a device that might catch fire, software flaws are easy to hide—until they aren't. The lack of immediate physical danger can lull organizations into complacency. But make no mistake: digital vulnerabilities can cause profound human harm. A data breach can ruin someone's privacy, finances, and even health (imagine medical records exposed or critical infrastructure hacked). The fact that your company handles sensitive customer data means you have an ethical and professional duty to protect it diligently. One of the core principles of engineering integrity is honesty and transparency. Ignoring a known flaw and not informing users or customers violates the principle of transparency. It's a lie by omission. You might not be telling a falsehood outright, but you're creating a false sense of security. That is unethical.

Let's put it in concrete terms. You said the product's vulnerability could allow an attacker to access user account information. Consider a worst-case scenario: an attacker exploits it next month after the launch, steals the personal data of thousands of clients, or injects malicious content. The breach comes to light. What then? Your company will have to confess it knew about the flaw beforehand. The leadership's decision to bury it will not only look irresponsible—it may even be considered negligence. The fallout could include regulatory fines, lawsuits from customers, and irreparable damage to reputation. We've seen this play out in real life. Consider the 2017 Equifax breach: a known software vulnerability was left unpatched, resulting in the theft of 147 million people's personal information. The public was outraged not just at the breach, but that Equifax hadn't acted fast enough on a known issue. Or recall how "security through obscurity"—hoping no one finds a flaw—failed countless organizations. Silence and hope are not security strategies; they are ethical failures.

Now, I understand you're relatively junior and fear the consequences of defying your bosses. It's a tough spot. But part of ethical leadership (and yes, you can lead from any position) is sometimes having to deliver unwelcome news upward. It might help to know that the law and industry standards are increasingly on your side. In many jurisdictions, companies are obligated to exercise "reasonable security practices" and could be penalized for willful neglect of known vulnerabilities. But beyond that, appeal to your leaders' self-interest positively: frame fixing this vulnerability as an opportunity to demonstrate the company's commitment to security and customer trust. Perhaps they fear a delay—but a breach would be a far bigger setback. Emphasize that "our customers have entrusted us with their data, and we owe them a duty of care." This aligns with what I taught about the wider constituencies of a technologist—you serve not just your employer, but also society and the public who rely on your systems.

If direct reasoning doesn't work, consider the broader support system. You mentioned there is an internal ethics or compliance team. Seek their confidential advice. Sometimes a concern gains more weight when echoed through official channels. Document your findings and your attempts to escalate the issue. This isn't just self-protection (though it's wise); it also creates an audit trail that might spur action. Many large tech firms have anonymous reporting for exactly this scenario—use it if needed. You can say, "I'm not comfortable leaving this unaddressed. If we don't fix it now, I feel obligated to at least document the risk and inform higher management." That might jolt them because it implies liability.

I also want to address the personal angle. Your stance may make you unpopular internally for a while. You might get labeled as "the alarmist" or the "boy who cried wolf." But those who truly understand security will respect you. Let me share that I have sat in rooms with executives after a breach, and inevitably, someone asks, "Did anyone know about this beforehand?" Imagine if your memo is produced in that meeting, showing you did warn them. I guarantee that in that future moment, you'd be regarded as the one person who tried to save the company from itself. Sometimes leadership is about being willing to say what others won't. Yes, it can be lonely. But it is also how you distinguish yourself as someone who stands for something more than the path of least resistance.

You also asked how this aligns with your career—could pushing this get you fired, and was it worth it? I can't promise outcomes, but I can share this: In my four decades of experience, the people who quietly do the right thing build a reputation that opens more doors than it closes. Even if one company doesn't appreciate it, others will. The tech world, especially in security, is a small community. Being known as the engineer who will speak truth to power, who won't cut corners with security—that's a badge of honor. In contrast, going along with something you know is wrong will eat at you. It's hard to put a price on the peace of mind that you did your best to prevent harm.

I'm reminded of a question I often pose: "Are we acting with integrity, even when it's inconvenient?" This is one of those moments for you. It is inconvenient indeed to raise a hand and say "Stop" when everyone else is charging ahead. But integrity means doing it anyway. Remember, trust can be lost in an instant. Your company trusts your expertise as a security professional, and your customers are implicitly trusting that you've done your due diligence. If the worst happens, saying "I was just following orders" will ring hollow. As a security expert, you are the voice of conscience for systems that cannot speak for themselves.

One more thought: sometimes framing an issue as an ethical one can cause defensiveness ("Are you saying I'm unethical?!"). So tactically, you might frame it as a business risk decision with ethical implications. For example: "If we launch with this vulnerability, our risk of a breach is X%. If that happens, we project X in loss and legal exposure. The right thing to do—and the smart thing—is to fix it now. Our brand is about trust. Let's live up to that." By tying ethics to tangible outcomes, you give management multiple incentives to do the right thing.

In closing, I encourage you to stay strong and use the resources at your disposal. Whether it's a mentor in the company, the ethics hotline, or even external professional advice, don't carry the burden alone. In the cybersecurity realm, the mantra is "see something, say something," just as much as it is in the physical security world. You have seen something; now you must say it clearly. However this turns out, you will know you stood on the side of integrity. Years from now that will matter more than any short-term fallout. And if you need further counsel or just a sympathetic ear, I'm here—as are your colleagues from the MSST program who have doubtless faced similar trials. You are part of a community of professionals dedicated to securing the world and making it better. We've got your back.

Sincerely,

Massoud

Letter to an Engineer Under Pressure

Dear Engineer,

You wrote to me about a situation that feels like a page out of a case study—only it's happening to you in real life. You're a young design engineer at a multinational automotive company. You've been working on a new electric vehicle design, specifically on the battery system. Lately, you've grown concerned that in rare stress conditions, the battery cooling might fail, risking a fire. You raised this internally, and the technical team acknowledged it as a low-probability risk. But when the issue traveled up the management chain, you were urged to "keep it quiet." The vehicle is slated for production, and any redesign would be costly, delaying the launch and affecting stock prices and executive bonuses. One manager even hinted that bringing this up again could affect your performance review. You're asking: Do I let it go, assuming the odds of failure are minimal? Or do I press the issue and potentially sabotage my standing at the company?

I can almost feel the weight on your shoulders. This is a classic ethical crunch in engineering—when business pressure collides with safety concerns. It resonates with the infamous story of the Challenger shuttle disaster. In that 1986 tragedy, some engineers warned that the O-ring seals in the rocket boosters could fail in cold weather. They begged management to delay the launch. Management, under pressure to stay on schedule, disregarded the warnings. Investigations later revealed that the voices of concern were stifled, and the decision-makers convinced themselves the risk was acceptably low. It was a catastrophic lesson in the cost of silence and overconfidence.

Your situation, thankfully, has not resulted in disaster—and with your action, hopefully never will. But it certainly is reminiscent of Challenger, and with other cases like the GM ignition switch scandal in 2014, where a known flaw in cars caused fatal accidents but was kept quiet for years. In these events, engineers or mid-level managers often knew something was wrong, yet the corporate inertia or fear prevented timely action. Eventually, the truth surfaced, as it tends to. Careers were ruined, companies paid dearly in recalls and lawsuits, and worst of all, people died needlessly. The pattern is chillingly consistent: someone raised the red flag and was ignored or pressured to hush up.

I don't recount these to scare you, but to reinforce that your instincts are right. If your gut is telling you something that could endanger lives, you must pursue it. As the saying in medicine goes, "First, do no harm." Engineering has a similar prime directive: protect the public. It's enshrined in every code of ethics for engineers. No probability is "too low" when human life is at stake. You must assume that if a failure is possible, it eventually will happen, and likely at the worst time. So what to do? You've already followed the first steps: you identified the issue and reported it internally. That took courage. Now, despite facing pushback, it's time to consider escalation paths.

Every reputable engineering organization (especially in the automotive industry, which is highly regulated) has channels for safety concerns. If your immediate superiors are unreceptive, you may need to involve a product safety board or ethics committee within the company. Many firms have a formal review process if an engineer flags a potential safety defect—precisely to avoid the scenario where business objectives steamroll safety. Use those mechanisms. It might feel like going over someone's head (and it is), but that is justified when those in charge aren't listening. Document everything: when you found the issue, who you told, and what their responses were. This isn't about building a legal case (though it incidentally does that); it's about making sure the concern is crystal clear and cannot be handwaved away as a "minor memo." If your company has a whistleblower or anonymous reporting system, consider using it if direct escalation could blow back on you severely.

It's also worth trying one more round of direct conversation, but framed differently. Instead of saying "this could cause a fire, we need to redesign," which they heard and resisted, reframe your concerns about the cost of not acting. For instance: "If we have a battery fire incident in the field, and it's traced back to this known issue, the recall costs and brand damage will dwarf the redesign cost now." In other words, speak their language. Management thinks in dollars and risk to the business. Show them that addressing the safety risk is not only ethically right but financially prudent. This approach isn't guaranteed to sway them, but it often garners attention. You can back it up with examples: "When Boeing ignored early warnings about MCAS, it later faced an estimated $20 billion hit after the crashes. When GM dragged its feet on the ignition switch issue, it paid for it in billions and reputational loss. Do we want to be on that list? Our company's reputation for safety is our most valuable asset." Sometimes putting it in stark, self-interested terms can break through denial.

Now, let's talk about you. I sense your fear—fear of retaliation, of derailing your promising career. This is understandable. You shouldn't have to risk your job to do the right thing, but history tells us that sometimes that's the price of admission to the Integrity Hall of Fame. However, before it comes to an ultimatum, remember that you do have allies. Are there other engineers or team members who agree with you? A united front is harder to shoot down. If a few of you jointly raise the concern, it carries more weight and protects individuals from being singled out. Also, do not underestimate the power of mentors. Is there someone higher up, maybe outside your chain of command, who has shown a commitment to ethics and might advise or champion this? Every company has a few senior figures known for their principled stance. Approach them discreetly for guidance.

You mentioned the hint about your performance review being impacted. That, frankly, is a form of intimidation and is inappropriate. If things turn ugly and you suspect actual retaliation, keep records and know that labor laws protect whistleblowers, especially on safety matters. It might never come to that, but knowing your rights can steel your resolve.

The Integrity Compass

Beyond the mechanics of reporting and persuading, I want to address the emotional aspect. Standing up in a pressure cooker environment tests what you're made of. Here's what I know about you (and those like you): you became an engineer to solve problems and help people. You're not in it just for a paycheck; you have a spark of idealism. That's evident because if you didn't, you wouldn't be agonizing over this—you'd just shrug and go along. That idealism is a strength, not a weakness. It will guide you well if you honor it. Many years ago, one of my mentors told me, "Your integrity is the only thing that can't be taken from you; you can only give it away." I never forgot that. No matter what the higher-ups do, they can't force you to be unethical; they can only pressure you to choose it. So, decide now that you won't give it away. There might be a price, but your integrity is worth it.

In closing, I support you fully in pursuing this issue. Remember the phrase "silent justice" we talked about in class: the idea that integrity is often quiet, shown not in grand gestures but in steadfast daily discipline. This is a moment of silent justice. It won't make you a hero in headlines, but it is heroism nonetheless. The people who will never know that their car is safer, their battery won't catch fire—they won't be able to thank you, but you will have done right by them. And that is the essence of engineering ethics: doing right by people who may never realize the care you took on their behalf.

Stay strong. Your future self—and countless others—will thank you for it.

Sincerely,

Massoud

Afterword – A Final Call to Action

One hero alone does not exercise ethical systems leadership at the top—it is a collective effort spanning all roles. Each of us in our spheres can practice it. A junior engineer who raises a concern in a meeting is exercising a form of leadership by influencing the system's trajectory. A mid-level manager who mentors their team in ethical thinking, a community member who provides constructive feedback to a local project, an academic who trains future engineers to value ethics—all are part of the mosaic of leadership that steers society's systems.

This collective nature is actually empowering: it means you do not have to wait to be a CEO or agency head to make a difference. By applying the ideas of this book in your current context, you are already contributing to ethical leadership. If you're a student, perhaps you can start an ethics discussion group or propose a code of conduct in your project team. If you're in the industry, you might gently educate peers when you see ethical blind spots, or volunteer to draft an ethics checklist for your group. If enough individuals do this, the culture shifts.

We also need to support one another in this journey. Ethical leadership can be lonely if one feels like the only voice of dissent.

But often, others share your concerns quietly. By speaking up, you encourage them to join. Find allies—maybe another colleague also worries about the rushed testing; team up to approach management. Networks and professional societies dedicated to ethics can provide solidarity and resources. For instance, groups like IEEE's Society on Social Implications of Technology or the Online Ethics Center offer forums and materials that can bolster your efforts and resolve ethical dilemmas.

On a broader level, part of ethical leadership now is advocating for structural changes that embed ethics, such as better regulations, standards, and educational curricula. Tech leaders should welcome sensible regulations that set ethical baselines (like data protection laws or AI bias requirements), rather than fighting all regulations, because good regulations level the playing field and prevent races to the bottom. Similarly, pushing for ethics and systems thinking in university programs will produce new generations of engineers ready to continue this work. My career involved teaching thousands of students and emphasizing ethics and systems thinking, thereby multiplying impact. In your own way, you can mentor or support education—maybe guest lecture at a local school about what ethical engineering means in practice, inspiring youth.

In concluding this book, it's fitting to return to hope and the belief that positive change is possible. Yes, we face daunting systemic problems: a warming planet, disruptive technologies outpacing policy, and deep social inequities. But as we've seen, ethical and inspired leadership *has* made a difference throughout history and can continue to do so even in our present day. It was ethical leadership that led to international agreements reducing ozone-depleting chemicals, that guided the development of the internet as an open platform (at least originally), and that underpinned efforts like the Green Revolution in agriculture to alleviate hunger. Often, these successes are not celebrated enough. Still, they show that with clarity of purpose and collaboration, we can "bend the arc" toward justice and sustainability.

Your role in this is both unique and shared. Your compass is yours alone—you must calibrate it and follow it. No one else can dictate your integrity. But you are not alone in using it: you are joining a community of engineers, scientists, thinkers, and citizens who are committed to building systems that uplift rather than exploit. When you face an ethical quandary and choose the principled path, know that around the world, many others are doing the same in their context, even if it's not making headlines. And those cumulative actions do add up.

This is a call to action: carry forward the lessons and stories you have encountered. When you design or lead, recall the ethical frameworks, ask the tough questions, engage those affected, and don't shy away from emphasizing values alongside analytics. In meetings, be the one who asks, "But who are we leaving out?" or "What if our assumption is wrong and something bad happens—do we have a backup?" Those questions can change outcomes. Encourage a culture of openness and care in your teams. Celebrate not just the clever solution, but the one achieved the right way.

And most importantly, take heart that striving for ethics is not about perfection—it's about direction and resilience. You will make mistakes; we all do. Ethical leadership is shown in how you respond: by acknowledging missteps, learning, and improving. It's like engineering itself—iterative. If you keep at it with humility and determination, you will get closer to the ideal over time. People will see that sincerity and respond in kind.

In closing, recall the image from the introduction: a compass pointing to true north. The world around us is changing rapidly—the terrain is shifting, storms will come and go—but true north (our core ethical values) remains steady. If you've tuned your compass well and practice using it, you'll navigate even uncharted territory with confidence. As an ethical systems leader, you won't just react to the future; you'll help shape it—creating technologies and institutions that are worthy of the trust we place in them. That is a legacy to be proud of. The next era of innovation and society are indeed in your hands, and with your ethical compass, I believe it will be a legacy of wisdom, inclusion, and hope. As I told my students at the end of every class, "May you always lead with integrity."

Acknowledgements

This book is dedicated to my late parents, Dr. Mohammad Shafi Amin (1914–1991) and Mrs. Nahid Loghman-Adham Amin (1924–2013), who raised me in an ethical home while storms raged outside.

To my mentors, and to the students and executives I have worked with during my thirty-eight years of teaching and leadership development. You entrusted me with your real ethical dilemmas, and in doing so, shaped the thinking and lived experience reflected in these pages.

With deep thanks to my editor, Ms. Emily Trenholm, at Calumet Editions, whose clarity, care, and discipline strengthened this work at every step.

Author Bio

Massoud Amin is a professor emeritus at the University of Minnesota, where he previously served as the Honeywell/H.W. Sweatt Chair and Director of Technological Leadership. He previously led critical infrastructure security R&D for North America's utility companies post-9/11 and has been an advisor for the White House, Congress, US agencies, and national laboratories. He serves as the Chief Technology Officer of Renewable Energy Partners and Chairman and President of Energy Policy & Security Associates.

Appendix A - Toolkit

Principles and Heuristics for Ethical Decisions

Drawing on my teachings and the cases we've explored in this book, here is a concise toolkit of principles and heuristics to guide you when facing ethical decisions in engineering and leadership:

- **Prioritize Safety and Welfare:** Always place the health and safety of the public first, ahead of schedules or profits. This is the first filter for any decision—if an action might cause undue risk to people or the environment, reconsider it or find a safer alternative. No deadline justification can excuse endangering lives. As a heuristic: If you wouldn't feel comfortable explaining a decision involving safety to a victim's family, don't do it.

- **Uphold Honesty and Integrity:** Be truthful in data, reports, and communications, even when the truth is uncomfortable. Integrity means no half-truths or hiding critical information. Practice transparency: share relevant facts with stakeholders and do not mislead by omission. A good rule of thumb: If you feel tempted to hide something, that "something" likely needs to be addressed, not hidden.

- **Accountability and Courage:** Accept personal responsibility for your work and its outcomes. Don't pass the buck. Remember that as a professional, your name is tied to your work—protect your good name by acting with integrity. This also means having the courage to say "no" or to pause a project when ethical lines are at stake. A mental check: Would I attach my name publicly to this decision? If not, it's a sign something's off.

- **Consult and Collaborate:** When in doubt, seek counsel from trusted colleagues, mentors, or ethics officers. Ethical decision-making is not a solitary endeavor; it involves diverse perspectives. Often, a quick conversation can illuminate blind spots or provide moral support for a tough call. Collaboration in ethics ensures you're not operating in an echo chamber of one. As a heuristic: If a decision has big implications, make sure more than one set of eyes (and values) has examined it.

- **Think Long-Term and Systemically:** Look beyond immediate gains to long-term impacts. Ethical engineering is about sustainability—will this decision hold up over time without causing harm? Consider the broader system: how does this choice affect not just direct users, but the community, environment, and future generations? Ask yourself: What story will be told about this project in five or ten years? Aim for one you'd be proud of.

- **Respect Regulations and Standards but Go Beyond:** Compliance with laws and standards is a baseline, not a ceiling. Stay well-informed of the regulations in your field (safety codes, privacy laws, etc.) and honor them. But true integrity might mean exceeding the minimum requirements when needed. "Legal" doesn't always mean "ethical." Use laws as guidance, but also apply your moral common sense and the principles listed here to fill in the gaps.

- **Document Deliberately:** Maintain clear documentation of key decisions, rationales, and data. Not only is this good engineering practice, but it is also an ethical practice—it keeps you honest and can serve as evidence of your thought process if scrutinized. It forces you to articulate why you chose a path, which often reveals if that choice was ethically sound or not (if you're uneasy writing it down, that's a telling sign).

- **Learn and Improve Continually:** Ethical decision-making skills grow with experience and reflection. Treat near-misses or smaller ethical dilemmas as learning opportunities. If you realize a past decision was not ideal, acknowledge it, learn why, and adjust your approach next time. An ethical leader is not one who never makes mistakes, but one who evolves. Build an internal feedback loop: after a project, ask "Did we uphold our values? Where could we do better?"

These heuristics can serve as a moral compass in daily work. They are straightforward by design—much like a pre-flight checklist for a pilot. Amid a complex project, revisiting these basics can help keep you oriented.

Questions for Maintaining Integrity

When facing a tough decision, asking the right questions can illuminate the ethical path. I often encouraged young professionals to pose a series of integrity-check questions. Here are some critical ones to incorporate into your decision-making process:

- **Impact on Stakeholders:** "How will our decision impact society and future generations?" Consider who is affected—not just directly, but indirectly. Are you considering the most vulnerable users? Who might be harmed, and who benefits?

- **Integrity of Action:** "Are we acting with integrity, even when it's inconvenient?" Strip away the excuses and pressures—is the choice consistent with your core values and principles? Would you do this in the same way if you had no external pressure?

- **Visibility and Transparency:** "How would this decision look if made public?" Similar to a common ethics test (the newspaper headline test), imagine explaining your action on the evening news or to a respected elder. If you'd be uncomfortable or defensive, that's a red flag.

- **Legal and Policy Alignment:** "Is the action legal and aligned with our organizational values?" This ensures baseline compliance. If it's illegal, it's off the table. If it clashes with stated company values or professional codes, why is that, and is there a justified reason or just a convenient one?

- **Alternatives and Creativity:** "Have we considered all our options, including the option to do nothing or a third way?" Ethical dilemmas often present false dichotomies ("either we ship the faulty product or we miss the deadline"). Step back and brainstorm—there may be a creative solution that mitigates harm and addresses the concern. Don't get trapped in an A vs B choice if both feel wrong; find C.

- **Consultation:** "Who should we talk to before finalizing this decision?" Is there an expert, an ethics board, legal counsel, or even representatives of the public who can give input? Sometimes a quick consultation can surface issues you didn't see.

- **Long-Term Reputation:** "What legacy will this decision build?" Think of your personal and your organization's reputation. Each action accumulates into a track record. Are you building a legacy of trust and respect, or might this be a story you hope people forget?

By systematically asking these questions, you create a kind of checklist for your conscience. They help pause the frenzy of "get it done now" and inject a moment of reflection. In practice, even a minute or two of such thoughtful questioning can prevent rash moves. For example, in the earlier letter about cybersecurity, asking "How would this look if the public knew we ignored the bug?" quickly clarifies the answer—it would look irresponsible at best. Or asking the medtech startup team "Is this aligned with why we founded this company?" promptly re-centers the discussion on patient welfare, not investor impatience.

One particularly powerful question I emphasize is: "What if everyone in my position did the same thing?" This is essentially Kant's categorical imperative in plainer terms. If every engineer hid flaws, where would we be? If every researcher fudged a little data, what would science become? Conversely, if everyone took a safety stand, how many disasters might be averted? Thinking in these universal terms helps break the spell of self-justification and highlights the broader ethical principle at stake.

Keep these questions visible—on a sticky note by your monitor, or a card in your notebook. They are your ethical litmus tests. Over time, you'll find that you can internalize them and quickly sense an answer. Early in your career, it's wise to consciously walk through them, especially in unfamiliar or high-pressure situations.

Practice Scenarios and Continuing the Conversation

Ethical skills, like technical skills, are honed by practice. Engaging with scenarios—whether real cases or hypothetical ones—can prepare you for when you face similar situations firsthand. My workshops at the University of Minnesota often used case studies and role-playing to simulate dilemmas. Here are a few scenarios drawn from those modules and beyond, offered as prompts for your reflection or discussion with peers:

- **The Whistleblower's Dilemma:** A researcher, Dr. LT, develops a food safety model that reveals a major food supplier's practices could shorten shelf life and risk spoilage. The company sponsoring part of the research pressures Dr. LT to bury the results, noting there's "no regulation forcing disclosure." Does the researcher stick to the data and publish, possibly angering the sponsor and risking funding? (This scenario mirrors real cases where industry-funded research tested scientists' integrity. Key question: Do you let truth speak, or do you yield to the golden handcuffs of funding?)

- **Conflict of Loyalty:** Terry is a grad student with a technique that could help another lab's project. Terry's advisor, however, has a private consulting conflict and forbids sharing the knowledge, citing competitive interests. Terry is torn between loyalty to his advisor and the broader collaborative spirit of science. What should Terry do? Discuss the balance between mentorship obligations and the free exchange of knowledge. (Realize the deeper issue: personal interests of seniors should not impede scientific progress or a student's sense of right.)

- **Design by Committee or Conscience:** You are part of an engineering team designing an AI feature for social media that can subtly manipulate user content to increase engagement. It technically works and will likely boost revenue, but you feel uneasy as it could exploit vulnerable users or worsen screen addiction. In team meetings, these ethical concerns are often dismissed with the rationale that "everyone in the industry does it." Do you push the issue, and how? (This scenario probes the oft-seen rationalization "if we don't do it, someone else will," and challenges you to find your stance on user well-being vs. business as usual.)

- **Cross-Cultural Quandary:** Your firm is deploying a water purification system in a developing region. The local partner suggests bypassing some safety tests to speed up installation, hinting that in their country, regulations are more "flexible" and that delays are costing lives (people need clean water now). You suspect the shortcuts could risk system reliability. How do you weigh the urgent humanitarian need against the risk of a sub-standard deployment? (This scenario forces consideration of context—sometimes ethics involves nuance, like balancing immediate needs vs. long-term safety, especially in cross-cultural or resource-limited settings.)

By wrestling with scenarios like these, either in a group or privately, you build your moral "muscle memory." You'll find patterns: often it comes down to courage, communication, and clarity of values. The more you practice thinking or talking through tough cases, the more confidence you'll have when a real one comes your way. It's similar to emergency drills—you hope never to face the fire, but you practice the escape route regardless.

Another practical exercise I recommend is the "ten years test": When stuck, imagine explaining your decision to a young engineer ten years from now. Would you frame it as a lesson in how to do the right thing, or a cautionary tale of a mistake to avoid? If the latter, reconsider your approach now while you still can.

As you go forward in your career, refer back to this toolkit whenever you feel that knot in your stomach signaling an ethical test. Use the principles as your compass, the questions as your map checkpoints, and the scenarios as your training ground. And remember, the goal of this primer has been to be clean, honest, and practical—ethics grounded in real life, not abstract theory. The real world will continue to evolve with new technologies, new pressures, and new dilemmas. But the foundational values—honesty, responsibility, respect for others, courage—endure. They will be your anchor in any storm.

Appendix B - Echoes from Other Minds

Love, Emotion, and Human Connection

"The most important thing in communication is hearing what isn't said."
Peter F. Drucker

"Love does not consist in gazing at each other, but in looking outward together in the same direction."
Antoine de Saint-Exupéry

"Not everything that is faced can be changed, but nothing can be changed until it is faced."
James Baldwin

"In the beginner's mind, there are many possibilities, but in the expert's, there are few."
Shunryu Suzuki

"Don't believe everything you think."
Byron Katie

"The limits of my language mean the limits of my world."
Ludwig Wittgenstein

War, Collapse, and Fragility

"Each time history repeats itself, the price goes up."
Ronald Wright

"An invasion of armies can be resisted, but not an idea whose time has come."
Victor Hugo

"War is what happens when language fails."
Margaret Atwood

"There are decades where nothing happens, and there are weeks where decades happen."
Attributed to Vladimir Lenin

"You may not be interested in war, but war is interested in you."
Attributed to Leon Trotsky

Systems, Feedback, and Complexity

"Every system is perfectly designed to get the results it gets."
Attributed to W. Edwards Deming

"When we try to pick out anything by itself, we find it hitched to everything else in the Universe."
John Muir

"The map is not the territory."
Alfred Korzybski

"The greatest shortcoming of the human race is our inability to understand the exponential function."
Albert A. Bartlett

"Nature does not hurry, yet everything is accomplished."
Lao Tzu

"What gets measured gets managed."
Peter F. Drucker

"If you can't describe what you are doing as a process, you don't know what you're doing."
W. Edwards Deming

Foresight, Timing, and Planning

"The time to repair the roof is when the sun is shining."
John F. Kennedy

"Chance favors only the prepared mind."
Louis Pasteur

"A problem well stated is a problem half solved."
Charles Kettering

"Prediction is very difficult, especially if it's about the future."
Attributed to Niels Bohr

"Between stimulus and response, there is a space. In that space is our power to choose our response."
Attributed to Viktor E. Frankl

"Vision without action is a daydream. Action without vision is a nightmare."

Japanese proverb

"The best way to predict the future is to create it."

Attributed to Alan Kay

"Intelligence is the ability to adapt to change."

Stephen Hawking

"As for the future, your task is not to foresee it, but to enable it."

Attributed to Antoine de Saint-Exupéry

Leadership, Power, and Responsibility

"Control is not leadership; management is not leadership; leadership is leadership."

Dee Hock

"To govern is to foresee."

Émile de Girardin

"We do not inherit the earth from our ancestors; we borrow it from our children."

Attributed to Chief Seattle

"You cannot swim for new horizons until you dare to lose sight of the shore."

William Faulkner

"Wisdom is knowing what to do next; virtue is doing it."

David Starr Jordan

Culture, Emergence, and Storytelling

"We shape our tools and thereafter our tools shape us."
Marshall McLuhan

"Culture eats strategy for breakfast."
Attributed to Peter F. Drucker

"The real voyage of discovery consists not in seeking new landscapes, but in having new eyes."
Marcel Proust

"History doesn't repeat itself, but it often rhymes."
Attributed to Mark Twain

"Sometimes when you're in a dark place, you think you've been buried, but you've been planted."
Christine Caine

"Myths are public dreams; dreams are private myths."
Joseph Campbell

"If you want to go fast, go alone. If you want to go far, go together."
Most commonly attributed as an African proverb

Ethics, Meaning, and Collective Choice

"We are caught in an inescapable network of mutuality, tied in a single garment of destiny."
Martin Luther King Jr.

"It is no measure of health to be well adjusted to a profoundly sick society."

Jiddu Krishnamurti

"Ethics is nothing else than reverence for life."

Albert Schweitzer

"The arc of the moral universe is long, but it bends toward justice."

Theodore Parker, popularized by Martin Luther King Jr.

"You never change things by fighting the existing reality. To change something, build a new model that makes the old one obsolete."

Buckminster Fuller

"Hope is not a lottery ticket you can sit on the sofa and clutch, feeling lucky. Hope is an ax you break down doors with."

Rebecca Solnit

"The only thing necessary for the triumph of evil is for good men to do nothing."

Attributed to Edmund Burke

"We must cultivate our own garden."

Voltaire

"The best criticism of the bad is the practice of the better."

Richard Rohr

"There is no such thing as an apolitical stance. Not acting is a decision. Not choosing is a choice."

Anonymous

"No one is too small to make a difference."

Greta Thunberg

Appendix C − Decision-Making Checklist

This checklist synthesizes guidance from responsible conduct of research (RCR), medical device security technologies (MSST), and industry ethics training modules used in our senior design course. Use it to evaluate any decision that has potential ethical consequences.

- **Identify the ethical question.** What is the key issue, and why is it a question at all?

- **Gather the facts.** What is known, what is uncertain, and what assumptions are being made?

- **Determine stakeholders.** Who will be harmed or benefited by each option? Include hidden or future stakeholders.

- **Check legality and policies.** Is the action legal? Does it comply with your organization's policies and professional codes (e.g., IEEE, NSPE)?

- **Test against core values.** Does the action align with your values and your organization's values? If you took this action, would you feel proud to explain it to your family or see it on the front page?

- **Explore alternatives.** Are there creative options that reduce harm or risk? Generate at least two viable choices.
- **Weigh consequences.** What are the short- and long-term impacts of each option on safety, equity, sustainability, and trust? Who bears the risk and who gains the benefit?
- **Seek counsel.** Discuss the dilemma with colleagues, mentors, legal counsel, human resources, or an ethics office. Diverse perspectives illuminate blind spots.
- **Decide and act transparently.** Document your reasoning, communicate your decision openly, and be prepared to adapt if circumstances change.
- **Reflect and learn.** After acting, reflect on the outcome. Did you uphold your principles? How can the process be improved?

Appendix D - Suggested Readings

To deepen your understanding of ethics, resilience, and systems thinking, consider these resources:

- **Codes of Ethics:** IEEE Code of Ethics; NSPE Code of Ethics for Engineers; ACM Code of Ethics. Read online and discuss with peers.

- **Responsible Conduct:** The National Academies of Sciences, Engineering, and Medicine, *Fostering Integrity in Research* (2017) and *On Being a Scientist: A Guide to Responsible Conduct in Research* (3rd ed., 2009)

- **Ethical Theory:** David Ross, *The Right and the Good* (1930), for a concise introduction to deontological ethics.

- **Case Studies in Practice:** Barbara MacKinnon and Andrew Fiala, *Ethics: Theory and Contemporary Issues* (2014); Robert M. Veatch, Amy M. Haddad, & Dan C. English, *Case Studies in Biomedical Ethics: Decision-Making, Principles, and Cases* (second edition, 2014).

- **Engineering and Failure:** Henry Petroski, *To Engineer Is Human: The Role of Failure in Successful Design* (1985);

William Kamkwamba and Bryan Mealer, *The Boy Who Harnessed the Wind* (2009), the latter being a true story of innovation under resource constraints.

- **Technology and Society:** Solon Barocas, Moritz Hardt, and Arvind Narayanan, "Fairness and Machine Learning: Limitations and Opportunities" (2018); Daniel Kahneman, Olivier Sibony, and Cass R. Sunstein, *Noise: A Flaw in Human Judgment* (2021).

- **Additional Perspectives:** Tracy Kidder, *Mountains Beyond Mountains: The Quest of Dr. Paul Farmer, A Man Who Would Cure the World* (2003) for social justice; and ethicist Jonathan Haidt, *The Righteous Mind* (2012) for moral psychology insights.

Appendix E - Teaching Guide

This primer can anchor courses in engineering ethics, sustainability, systems thinking, and leadership. Suggestions for instructors:

- **Blend narrative and analysis.** Pair personal stories and letters with technical readings. For example, follow the letter to a medical innovator, which includes articles on device regulation and human subject protection.

- **Use frameworks repeatedly.** Assign the ethical decision-making checklist as a recurring tool. Ask students to document how they applied each step to a class project or current event.

- **Engage with case studies.** Discuss real examples drawn from the book and the workshop slides, such as the Boeing 737 MAX failures, Dan Markingson's clinical trial, and cybersecurity breaches. Encourage role playing: one team represents engineers, another investors, another regulators, and another impacted communities.

- **Practice Keep Asking.** Teach students the Keep Asking model from ethics training: Is the action legal? Does it align with our values? How will it look in the newspaper? If you know it's wrong, don't do it. If unsure, ask colleagues, legal counsel, or an ethics office.

- **Connect to systems thinking.** Link the systems chapters to quantitative modeling or simulation exercises. Students might model a supply chain shock, a tipping point in climate systems, or a social feedback loop.

- **Invite diverse voices.** Bring in guest speakers from industry, government, NGOs, and community organizations to discuss how they navigate ethical dilemmas in practice.

Appendix F - Discussion Prompts

- **Deadline versus safety:** Recall a moment when you faced a conflict between meeting a deadline and ensuring the quality or safety of your work. How did you decide? Would you act differently now?

- **Stories that resonate:** Which story or dispatch in this book resonated with you the most? Why?

- **Loyalty and the public good:** How do you balance loyalty to your employer with your responsibility to the public?

- **AI and automation:** In an era of AI and automation, are new ethical principles needed, or are existing ones sufficient? Discuss with examples.

- **Learning from failures:** After reading about the Boeing 737 MAX tragedies, how could earlier attention to ethical signals have prevented the disaster? What role did organizational culture play?

- **Reporting vulnerabilities:** Imagine you discover a security vulnerability in a widely used product. What steps do you take to report it? To whom?

- **Research ethics:** Consider a research misconduct case, such as the Dan Markingson clinical trial. What systemic failures allowed it? How might Responsible Conduct of Research practices prevent similar cases?

- **Unintended consequences:** Discuss a recent technology that has had unintended negative consequences. How could systems thinking have predicted or mitigated them?

- **Hope as a resource:** What does hope as a technical resource mean to you? How can it be cultivated in your community?

- **Equity in design:** How might the ethical considerations differ when designing technologies for affluent users versus marginalized communities? After reading this primer, what action will you take in the next month to strengthen your ethical practice?

Glossary - The Language of Systems

Actuator: A system component that directly changes an environment's physical state based on control signals. In engineered systems, this could be a motor; in human systems, it might be a decision or action.

Adaptation: The process by which a system modifies its internal structure or behavior in response to external change or internal feedback is a hallmark of resilient and intelligent systems.

Agent: An autonomous unit within a system that acts based on rules, incentives, or perceptions—e.g., a person, cell, sensor, or AI entity.

Algorithmic Bias: A distortion in outcomes caused by assumptions, training data, or design choices embedded in algorithmic systems.

Attractor: A state or set of states toward which a system tends to evolve. Can be stable, unstable, fixed, periodic, or chaotic.

Autonomy: A system or subsystem can operate independently and make decisions without external control.

Bifurcation: A qualitative change in a system's behavior caused by a small change in a parameter—a fork in the road of system dynamics.

Boundary Conditions: The constraints or limits that define the space within which a system operates. Changing these often alters system behavior.

Chaos: Apparent randomness arises from deterministic systems due to extreme sensitivity to initial conditions. Predictable in form, unpredictable in outcome.

Closed-Loop System: A system with feedback. Output is measured and compared to a desired input, and corrections are made accordingly.

Complex Adaptive System (CAS): A system composed of interacting agents that evolve. Examples include ecosystems, economies, organizations, and the brain.

Control: Control is an input or decision that steers a system toward a desired state. It can be centralized or distributed, explicit or emergent.

Coupling: The degree to which components of a system influence each other. Tight coupling means a rapid and significant impact; loose coupling allows for independence.

Critical Point: The threshold beyond which a system undergoes rapid change or collapse. Often linked with tipping points or phase transitions.

Damping: A mechanism that reduces oscillation or instability in a system. Analogous to braking or resistance.

Delay: The time lag between an input or action and its observable effect. If not accounted for, delays can destabilize systems.

Disturbance: An external shock or unexpected event that alters the state of a system is known as a perturbation. It can reveal hidden fragility or trigger adaptation.

Emergence: The appearance of patterns, structures, or behaviors at a higher level that are not evident from individual components. "The whole is more than the sum of its parts."

Equilibrium: A state where the system is stable and its variables do not change unless disturbed.

Feedback: Information about a system's output that is looped back into the system to influence future behavior. Can be positive (amplifying) or negative (stabilizing).

Fractal: A structure that exhibits self-similarity across scales. Used metaphorically to describe patterns in human behavior, markets, or nature.

Homeostasis: A system's tendency to maintain internal stability despite external changes is known as homeostasis.

Inertia: A system's resistance to change. Often driven by legacy processes, habits, or institutional memory.

Input: A signal, force, resource, or decision that enters a system and affects its behavior.

Instability: A condition where small disturbances grow over time, pushing the system away from equilibrium.

Interdependence: A condition where elements within a system rely on each other to function. Increases both complexity and fragility.

Latency: The time between the cause of an event and its observable outcome. Often critical in early warning systems.

Leverage Point: A place within a system where a small shift can lead to large changes in outcomes.

Loop: A circular path in which outputs affect future inputs. Can be reinforcing or balancing.

Markov Process: A random process where the next state depends only on the current state—memoryless transitions.

Model: A simplified representation of a system used to understand, predict, or influence behavior.

Node: A connection point in a networked system is a person, computer, intersection, or policy unit.

Nonlinearity: A property where the output is not proportional to the input. Small causes can have large effects, and vice versa.

Oscillation: A repeated fluctuation around a central state or value. This phenomenon is common in economic cycles, ecosystems, and emotional dynamics.

Output: The result or behavior produced by a system after processing inputs.

Overshoot: A condition where a system exceeds its limits, often leading to collapse or correction.

Parameter: A variable that characterizes system behavior but is not part of the state itself. Changing parameters can lead to bifurcations.

Path Dependence: A dynamic where past decisions or states constrain future possibilities.

Phase Shift: A transition in a system's behavior—from stability to chaos, or from one dominant pattern to another.

Resilience: The ability of a system to absorb shock, adapt, and continue functioning. Not the same as strength—resilience bends but doesn't break.

Robustness: The capacity of a system to maintain function despite disturbances or uncertainty.

Runaway Loop: A reinforcing feedback cycle that accelerates without internal braking—e.g., a bank run, a viral meme, or exponential love/hate dynamics.

Scenario Planning: A method for exploring possible futures by modeling different sets of inputs, choices, and constraints.

Sensor: A component that measures a specific variable and provides input to the system for feedback or decision-making.

Signal: A meaningful change in data or behavior that indicates system state, direction, or risk. Distinguishable from noise.

Slack: Unused capacity in a system that allows for adaptation under stress. Critical for resilience.

State Variable: A measurable quantity that describes the current condition of the system (e.g., voltage, mood, GDP, heart rate).

Stability: A system's ability to return to equilibrium after a disturbance.

Stochastic Process: A system or model driven by random probability rather than deterministic inputs.

System: A set of interacting elements organized to achieve a purpose. Systems have boundaries, inputs, outputs, feedback, and internal states.

System Drift: Gradual, often unnoticed, movement away from optimal or safe conditions. It is dangerous when combined with delay or blind spots.

Tipping Point: A critical threshold beyond which a small input causes irreversible system transformation.

Tradeoff: Optimization problems, ethics, and engineering involve situations in which improving one part of a system worsens another.

Unintended Consequences: Outcomes that are not foreseen or intended often result from system complexity or flawed models.

Utility Function: A formula or rule that expresses preferences and guides decisions within optimization models.

Variable: Any element of a system that can change over time. Variables are measured, monitored, or manipulated.

Vulnerability: The degree to which a system is susceptible to harm from disturbance or stress.

Window of Opportunity: A temporary period in which sytem change is possible before inertia, feedback, or delay closes it again.

www.ingramcontent.com/pod-product-compliance
Lightning Source LLC
Chambersburg PA
CBHW031611210526
45464CB00004B/1525